INTRODUCTION TO HOMELAND SECURITY

PREPARATION, THREATS, AND RESPONSE

By Mark M. Lanier and Edmund "Ted" Sexton

University of Alabama—Tuscaloosa

cognella®
academic publishing

Bassim Hamadeh, CEO and Publisher
Michael Simpson, Vice President of Acquisitions
Jamie Giganti, Managing Editor
Jess Busch, Graphic Design Supervisor
Zina Craft, Acquisitions Editor
Brian Fahey, Licensing Associate
Sean Adams, Interior Designer

First published in the United States of America in 2014 by Cognella, Inc.

Printed in the United States of America

ISBN: 978-1-62131-712-8 (pbk) / 978-1-62131-713-5 (br)

cognella®
academic publishing

www.cognella.com 800-200-3908

Contents

Section 3: Response

Section 1 PREPARATION

Chapter 1

Genesis, Overview, and Mission of the Department of Homeland Security

Joseph "Andy" Musick, Eldon T. Ferguson, and Mark M. Lanier,

Introduction

> Today's terrorist can strike at any place, at any time, and with a wide variety of weapons ... The terrorist threat to America takes many forms, has many places to hide, and is often invisible. We can never be sure that we have defeated all of our terrorist enemies, and therefore we can never again allow ourselves to become overconfident about the security of our homeland.
>
> —President George W. Bush, "Securing the Homeland, Strengthening the Nation," 2002

The Department of Homeland Security (DHS) was created in response to the September 11, 2001, terrorist attacks on the World Trade Center and the Pentagon. The United States found itself in a paradoxical situation when these attacks occurred. Despite enormous wealth and resources relative to its attackers, the nation found itself unprepared to prevent such attacks and coordinate responsive efforts to mitigate the damage (Schmitt and Shanker, 2011). As the United States prepared for war against Al Qaeda and the Taliban in Afghanistan, President George W. Bush and other lawmakers saw an opportunity to create a department that could use the nation's vast resources in coordinated efforts to aid in disaster relief.

Problems Prior to the Creation of DHS

The U.S. governmental structure prior to the 9/11 attacks was a mixture of convoluted responsibility and isolationist tendencies, in consequence with poor communication and little sharing of resources, data, and information. According to Wise and Nader (2001), there were two main problems. First, the United States inadequately used its existing resources due to a lack of coordination, coupled with overlapping jurisdictions. Second, state and local officials were not sure which federal agency or department authority (and/or command) in certain situations, in part due to communication issues and regional differences, were in charge.

Inadequate Use of Resources

The U.S. government's inability to use its resources to defend itself from terrorist attacks brought coordination and jurisdictional problems to the forefront. During this time, the United States was regarded as having the strongest military, the best intelligence services, most natural resources, and the largest economy. Despite having these resources, a few men were able to hijack planes using box cutters, and then used those planes as weapons to kill thousands of people. The attack on 9/11 demonstrated an inability to use the relative advantages the United States had to defend itself effectively.

Because of overlapping jurisdictions, the federal agencies did not know who was responsible in certain scenarios. Before the 9/11 attacks, what would become the DHS was divided across more than 22 federal agencies and 2,000 separate congressional appropriations accounts. Each of these agencies had its own unique purpose, but each also had one fundamental goal: protect the United States against hostile threats (Department of Homeland Security (DHS), 2008).

When the 9/11 terror attacks occurred, multiple federal agencies and departments used their respective resources to address a massive interdepartmental problem. What became readily apparent was that these agencies and departments were not working with each other to determine who was responsible and to mitigate the damage of the attack, much less prevent the attacks (Schmitt and Shanker, 2011). The coordination among agencies was limited, at times even nonexistent. Each agency or organization largely relied on its own communication systems and made little to no effort to reach out to the other organizations (Aid, 2009; Schmitt and Shanker, 2011).

Jurisdictional overlap presented another problem. For instance, the 9/11 attack was a transnational criminal incident that affected the critical infrastructure of the United States. Organizations such as the Federal Bureau of Investigation (FBI), the Central Intelligence Agency (CIA), the Secret Service (USSS), and many more agencies are all charged with protecting critical infrastructure (Schmitt and Shanker, 2011). The mission statements of organizations such as the FBI, CIA, and the USSS are similar and overlap to varying degrees. Despite being tasked with similar responsibilities, these organizations were either unwilling or unable to share intelligence with one another (Schmitt and Shanker, 2011). While this was recognized as a problem prior to 9/11, little had been done to solve the issue. The attacks on 9/11 prompted immediate action to address these long-existing problems.

Combining some of these agencies into one department had been discussed previously. Federal executives at the U.S. Commission on National Security/21st Century, also known as the Hart-Rudman Commission, suggested the possibility of combining the agencies that focus on the protection of the United States from internal and external threats in January 2001 (DHS, 2008). Later, in March 2001, Texas representative Mac Thornberry proposed a bill that would create a National Homeland Security

Agency that would combine FEMA, Customs, Border Patrol, and several infrastructure offices together and create one agency responsible for homeland security-related activities. The agency would establish communication throughout each of the supplementary agencies in order to create a more effective department, which would be able to respond more efficiently to national crises (DHS, 2008). Both efforts failed in establishing a new agency or department. However, the realization that the government failed to protect its citizens—both before and immediately after the attacks—despite enormous resources resonated with lawmakers and government officials (DHS, 2008).

Confusion of State and Local Officials

If the federal agencies were unsure of whose responsibility it was to take charge in emergencies, then the state and local officials were equally, if not more, confused. When an incident occurs within the borders of the United States, it is only natural to have state and local police, firefighters, and other authorities respond to the incident. If it becomes apparent that federal entities have jurisdiction, then state and local authorities must be subservient to those federal entities. Arguably, if there is confusion among those federal organizations as to who should be in charge of a given situation, then that confusion trickles down to the state and local levels (Wise and Nader, 2001). For instance, police helicopters flying overhead were broadcasting on their radios that the South Tower of the World Trade Center was going to collapse, but firefighters inside did not hear this transmission (Dwyer, Flynn, and Fessenden, 2002). As a result, many of those firefighters were killed. Clearly, if an issue of national security arises, local and state officials will not be adequately prepared to handle the situation in isolation, but if they are unable to discern who their superiors are, the situation will only worsen (Wise and Nader, 2001).

Regional differences such as terrain, local governmental systems, and specific situations were not always taken into account (Wise and Nader, 2001). First, the landscape itself can alter incident and disaster strategy for prevention and relief. Second, different regions and states have different governmental systems. There are differences when dealing with counties, municipalities, and towns that governmental agencies were sometimes unaware of (DHS, 2008). Finally, many regions of the United States are not equipped to handle certain situations that may arise; if there is a lack of coordination between local and federal agencies, problems are only exacerbated (Government Accountability Office (GAO), 2004).

Key Questions

Before the creation of the DHS, there were four fundamental questions that often went unanswered due to the aforementioned issues (Aid, 2009; Schmitt and Shanker, 2011). First, who should be in charge at the federal level, and what is the chain of command if multiple federal agencies are involved? Second, what should the role of state and local officials be? Third, what role do regional differences play? Finally, accounting for all of these factors, what is the best way to coordinate effective responses to emergency? Most of these questions did not have simple answers—the attack on September 11, 2001, made that clear.

Creation of the Department of Homeland Security

> *Our Nation learned a painful lesson on September 11. American soil is not immune to evil or cold-blooded enemies capable of acts of mass murder and terror ... A new wave of terrorism, involving new weapons, looms in America's future. It is a challenge unlike any ever faced by our nation. But we are not daunted. We possess the determination to defeat our enemies and to work to secure our homeland against the threats they pose.*
>
> —President George W. Bush, "Securing the Homeland, Strengthening the Nation," 2002

The Department of Homeland Security (DHS) was created through a series of legislative steps and processes. First, on September 21, 2001, President George W. Bush stated that the United States needed an Office of Homeland Security (OHS) in the White House to oversee terrorist activity, coordinate a national strategy that would better prepare and safeguard the United States against terrorism, and respond to future attacks (DHS, 2008). On October 8, 2001, President Bush addressed a joint session of Congress and issued Executive Order 13228, known also as Homeland Security Presidential Directive-1 (HSPD-1). The order established two new entities: the Office of Homeland Security (OHS) and the Homeland Security Council (HSC). The OHS, located within the Executive Office, would take charge of developing and implementing a national strategy. This strategy coordinated federal, state, and local counterterrorism efforts to prepare and respond to terrorist threats or attacks. The HSC is composed of Cabinet members responsible for homeland security–related matters. It is designed to advise the president on homeland security issues (DHS, 2008).

By February 2002, President Bush released the first fiscal year budget numbers occurring after the 9/11 attacks. For the fiscal year of 2003 (FY2003), $37.7 billion was appropriated for homeland security efforts. Specifically, the budget allocated more funding for first responders in emergency situations, bioterrorism prevention efforts, border security, and technology.

Next, on June 6, 2002, President Bush outlined four goals that detailed what he felt the department should accomplish (DHS, 2008). First, it should monitor and control national borders in order to prevent terrorists and explosive devices from entering the country. Second, the DHS needed to work with local and state officials in order to respond as quickly and efficiently as possible to emergency situations. Third, the department should develop technology that would aid in the detection and prevention of the use of chemical, biological, and nuclear weapons. Finally, the department should consolidate data and intelligence from all government agencies in order to create concise, accurate reports regarding the security of the nation (DHS, 2008).

Next, in July of 2002, the White House introduced the first National Strategy for Homeland Security (NSHS) (Bush, 2007). The NSHS focused on the prevention of terrorist activity, citing three main objectives. First, the United States should prevent terrorist attacks within its borders. Second, the nation needs to reduce its level of vulnerability to terrorism. Finally, if terror attacks occur, the damage must be minimized.

Finally, on November 25, 2002, President Bush signed the Homeland Security Act of 2002 into law. The act officially created the Department of Homeland Security (DHS). This department was designed to meet three goals: fulfill the objectives of the NSHS, ensure the economic security of the United States, and monitor connections between drug trafficking and terrorism (Homeland Security Act, 2002). DHS became officially operational on January 24, 2003. Pennsylvania governor Tom Ridge was sworn in as the first secretary of Homeland Security that same day.

Structure of the DHS

The department was created by consolidating other government agencies and their existing personnel. The figure below describes the location of these organizations prior to the DHS and how they were moved into the department upon its creation.

Each of these organizations was chosen due to its ability to meet one or more of the goals outlined in the Homeland Security Act of 2002. The figure below shows a general breakdown of the command and hierarchical structure of the DHS.

As the DHS was being conceptualized and actively created, President Bush issued Homeland Security Presidential Directives (HSPD) to guide its growth as an organization. In total, 25 HSPDs were issued, the last in 2009 (Homeland Security Digital Library). However, the first seven were instrumental in the creation of the DHS. The table below gives brief descriptions of each of these directives.

How Effective Is the Department of Homeland Security?

The DHS is young relative to other government agencies and is going through "organizational learning" (Wise and Nader, 2002, p. 55). Government departments that have been moved under the umbrella of Homeland Security are learning how to work together more efficiently and effectively. However, the department does not have the luxury of slowly learning how to be an effective organization due to the nature of its responsibilities. Because of this, the organization has had some successes and failures. To date, there have been a few incidents that have raised questions about its effectiveness. However, many DHS successes will not be noted by the media or general public. For example, a foiled terror attack is not likely to be widely disseminated to preserve and protect strategies and informants, etc.

Hurricane Katrina

The department has been widely criticized for its mishandling of Hurricane Katrina relief efforts (Waugh and Streib, 2006). The Federal Emergency Management Agency (FEMA) was moved to the DHS because of its responsibilities in disaster management and relief. FEMA was created by an executive order issued by Jimmy Carter in 1979. When the agency was moved, it became part of a department

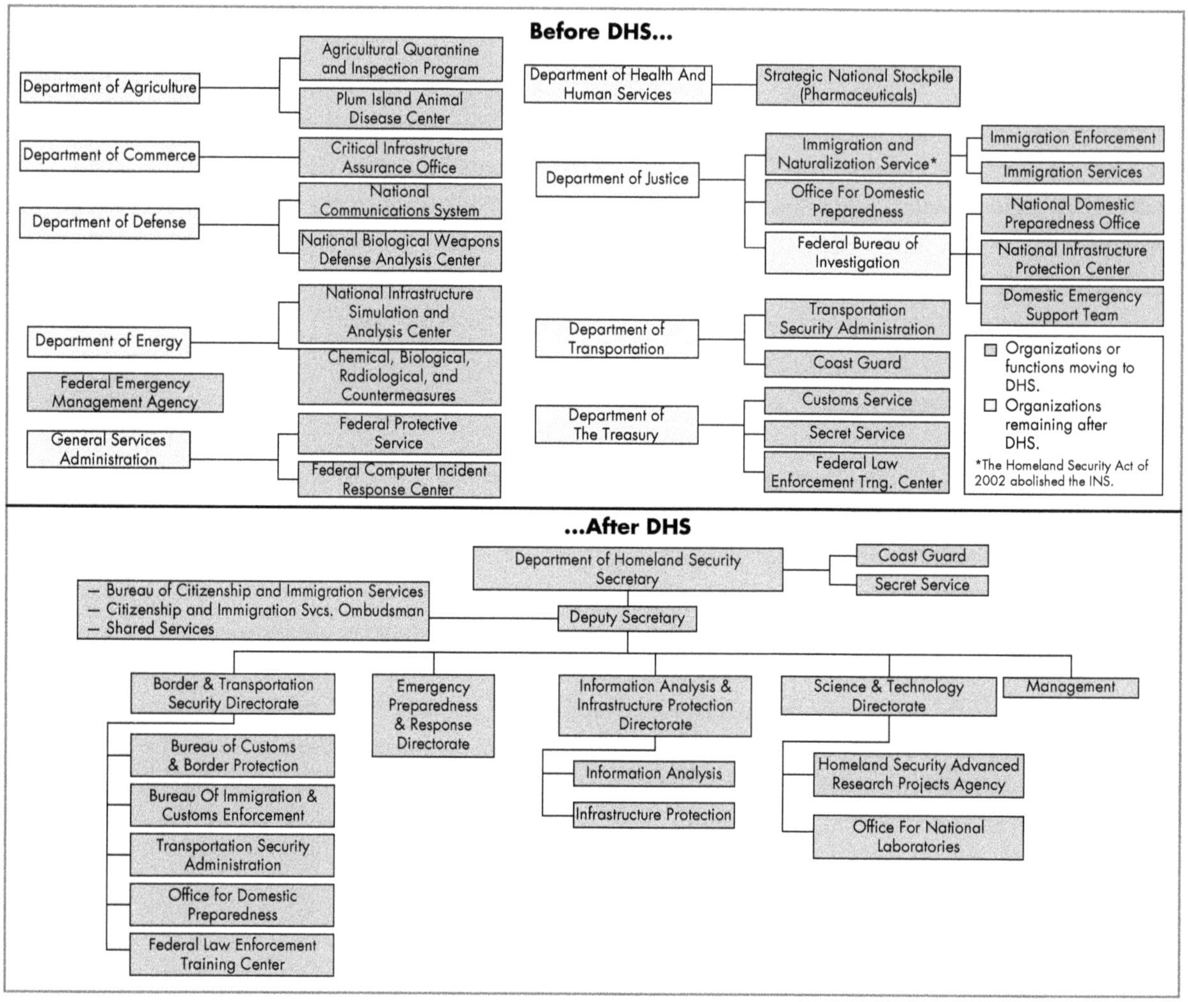

Source: U.S. Department of Homeland Security. Copyright in the Public Domain.

Table A1: Description of the First Seven Homeland Security Presidential Directives

Presidential Directive Number	Year	Name of Directive	Description of Directive
HSPD-1	2001	Organization and Operation of the Homeland Security Council	Ensures coordination of all homeland security-related activities among executive departments and agencies and promote the effective development and implementation of all homeland security policies.
HSPD-2	2001	Combating Terrorism Through Immigration Policies	Provides for the creation of a task force which will work aggressively to prevent aliens who engage in or support terrorist activity from entering the United States and to detain, prosecute, or deport any such aliens who are within the United States and to detain, prosecute, or deport any such.

continued

HSPD-3	2002	Homeland Security Advisory System	Establishes a comprehensive and effective means to disseminate information regarding the risk of terrorist acts to Federal, State, and local authorities and to the American people.
HSPD-4	2002	National Strategy to Combat Weapons of Mass Destruction	Applies new technologies, increased emphasis on intelligence collection and analysis, strengthens alliance relationships, and establishes new partnerships with former adversaries to counter this threat in all of its dimensions.
HSPD-5	2003	Management of Domestic Incidents	Enhances the ability of the United States to manage domestic incidents by establishing a single, comprehensive national incident management system (NIMS).
HSPD-6	2003	Integration and Use of Screening Information	Provides for the establishment of the Terrorist Threat Integration Center.
HSPD-7	2003	Critical Infrastructure Identification, Prioritization, and Protection	Establishes a national policy for Federal departments and agencies to identify and prioritize United States critical infrastructure and key resources and to protect them from terrorist attacks.

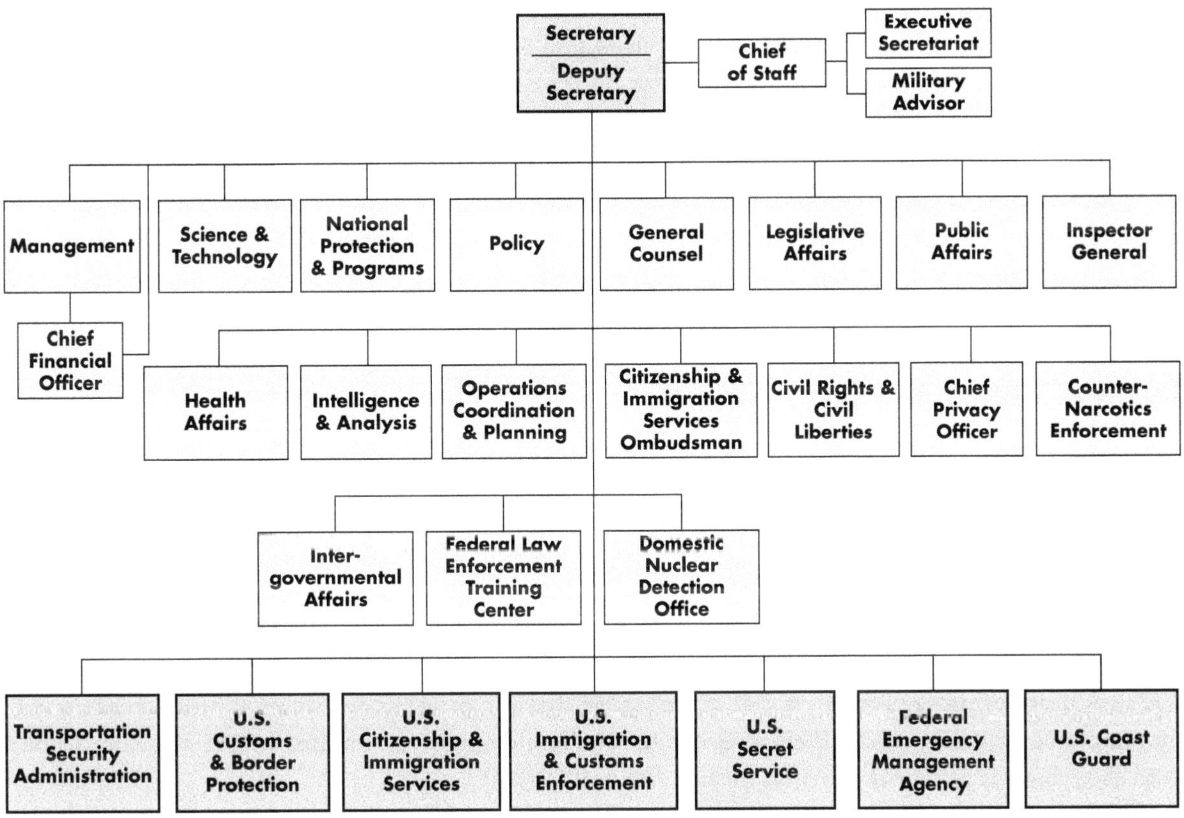

Source: U.S. Department of Homeland Security. Copyright in the Public Domain.

that was devoted to stopping terrorism and providing security. However, FEMA was originally created to address issues surrounding natural disasters. Many of the problems that occurred during the disaster relief efforts in New Orleans and the nearby coastal regions stemmed from the fact that the DHS was not focused on dealing with natural disasters, despite FEMA now being under its control (Birkland, 2009). Ironically, the DHS faced one of the very problems it set out to resolve: the assignment of responsibility and roles in disaster. Because FEMA was now within its organizational structure, the DHS encountered roles and responsibilities that were unforeseen, and it was fundamentally unprepared to handle a disaster of the magnitude of Hurricane Katrina (Cooper and Block, 2006).

Wastefulness

Second, there has been criticism of DHS effectiveness because of wastefulness. According to a report in the *Washington Post*, the DHS was in charge of $15 billion worth of failed projects and contracts (Hedgpeth, 2008). In 2006, a Government Accountability Office (GAO) report determined that "45 percent of DH's purchase card transactions were not properly authorized, 63 percent did not have evidence that the goods or services were received, and 53 percent did not give priority to designated improper transactions, which resulted in losses to the federal government (GAO, 2006). A *New York Times* report found that one FEMA employee spent "$7,790 on a 63-inch plasma monitor, which sat for six months, unused in its original carton" and another "spent $68,442 on the 2,000 dog boots, which were intended to protect the paws of search dogs on rescue operations" (Lipton, 2006). Regardless of the intent of the spending, this shows ineptitude on the part of the DHS to manage its funds and acquisitions correctly.

Fusion Centers and Data Mining

A third criticism stems from the DHS's use of fusion centers and data mining practices. A fusion center is a partnership between public and private entities to share information in order to aid in defense and prevent terrorism (Monahan, 2009). This type of information sharing is one of the goals that the DHS wanted to achieve. However, critics have pointed out flaws in the system. The GAO found the DHS had not set standards for the performance of such centers (GAO, 2010). Without clear guidelines, there is the possibility for wasted spending, privacy violations, and the overstepping of authority (Monahan, 2009). A more pressing problem is that it is possible that the private companies in fusion centers may have a desire to influence the DHS in order to label competitors or entities whose motives conflict with their own as potential terrorist threats (McAuliff, 2012).

Data mining is one of the tools the DHS uses to monitor terrorist activities. It is useful for looking at patterns involving monetary transfers, communications, and tracking terrorist movements (GAO, 2011). Advocates of the practice point out that it can extract useful information from vast amounts of data that otherwise would be difficult to pinpoint (Witten, Frank, and Hall, 2005). The use of data mining in the intelligence community has led to military victories, arrests of criminals and terrorists, and helpful strategic analysis (Aid, 2009; Schmitt and Shanker, 2011). The information gathered from the practice has proved invaluable in aiding the DHS and the federal government as a whole.

However, the GAO report on the practice of data mining points out a few problems: the value of trends is sometimes difficult to identify, it is hard to establish causal relationships, the quality of the data can vary, different agencies use different software, using data for unintended purposes, and privacy. First, if trends are not analyzed in their proper context, they can lead to inaccurate results and conclusions (Larose, 2006). Second, establishing correlational relationships does not equate to finding causal relationships (Witten, Frank, and Hall, 2005). Third, combining data from more than one source can lead to problems such as loss of information and inaccurate records (Larose, 2006). Fourth, agencies using different software can create communication issues and slow the analysis process. Fifth, using data for unintended purposes can lead to legal problems and rights violations (Lanier and Briggs, 2013). Finally, there are privacy rights concerns. For example, one data mining program, ADVISE, was ended in 2007 due to heavy criticism from privacy rights advocates (Sniffen, 2007).

The advantages of using these practices keep them in place. Their ability to provide useful statistics and information allows for effective analysis of problems. However, it is important to be aware of their potential misuses.

NIMS, NRF, and NCCIC

To better coordinate with agencies and departments at all governmental levels, the National Incident Management System (NIMS) and the National Response Framework (NRF) were created. Through these two programs, the federal government has laid out guidelines for incident response, disaster recovery, and communication. NIMS is the system through which governmental employees communicate (DHS, 2008). The NRF has the guidelines that governmental employees are to follow when incidents occur (DHS, 2008). These two systems working in tandem allow for streamlined communication and more efficiency when responding to emergencies. Figure 3 below shows how information is exchanged among the different organizations.

The DHS is also in charge of the National Cybersecurity and Communications Integration Center (NCCIC). The goal of this center is to protect the critical infrastructure of the United States from cyber-attacks. The mission of the NCCIC is to be the leader in protecting "federal civilian agencies" in the cyber world. Second, it is to work with "critical infrastructure owners and operators" to help manage risk. Third, the NCCIC must collaborate with local and state governments through the Multi-State Information Sharing and Analysis Center (MS-ISAC) and engage in cooperation with international allies and partners to trade intelligence and respond to emergencies. Fourth, through the use of the

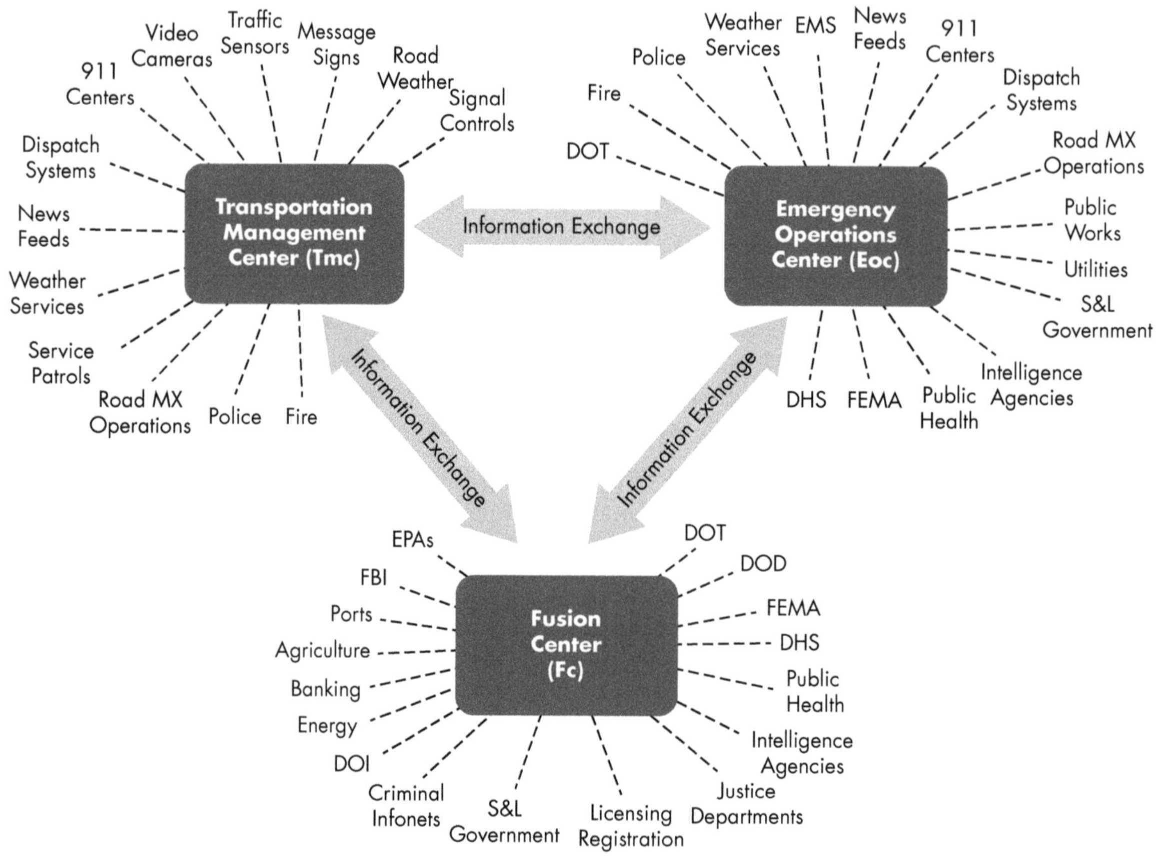

Source: U.S. Department of Transportation. Copyright in the Public Domain.

National Cyber Incident Response Plan, it addresses cyber-related problems that arise, analyzes data to better address cyber issues, creates and maintains "situational awareness," and orchestrates "national protection, prevention, mitigation, and recovery activities" related to cyber incidents. Finally, the NCCIC assists in "the initiation, coordination, restoration, and reconstitution of National Security or Emergency Preparedness (NS/EP) telecommunications services and facilities under all conditions, crises, or emergencies, including executing Emergency Support Function 2-Communications (ESF-2) responsibilities under the National Response Framework (NRF)" (DHS, 2013). Figure 4 (below) provides a breakdown of the DHS cyber-security command structure (DHS, 2013).

Terrorism Prevention

The general public will never know about many of the DHS's successful moments. Many media members are not privy to or concerned with successful terrorist prevention operations. This is likely due to the desire of the DHS to maintain a level of operational secrecy. If the media is not aware of successful

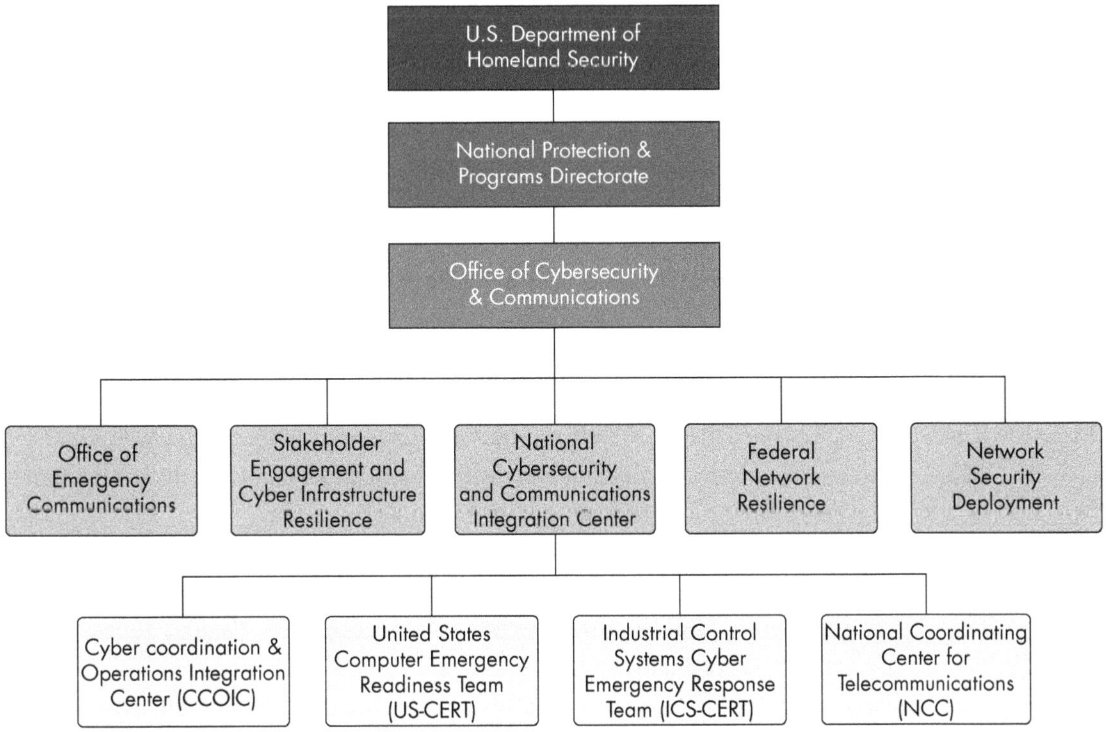

Source: U.S. Department of Homeland Security. Copyright in the Public Domain.

operations, then they cannot report on them. Additionally, if editors of media outlets do not find the story of a successful operation compelling enough, it is possible that the story will never be broadcast to the public.

To the Future

The DHS is an organization with the potential to greatly shape the future of the United States. In this book, more of these successes and failures will be analyzed to shed light on an important department that few truly understand. The Department of Homeland Security has numerous important roles that affect the lives of U.S. citizens on a daily basis. It is the goal of the authors to discuss in depth each of these roles, their importance, and what the future holds for the department.

References

Aid, M. M. (2009). *The secret sentry: The untold history of the National Security Agency*. New York: Bloomsbury Press.

Birkland, T. A. (2009). Disasters, catastrophes, and policy failure in the Homeland Security era. *Review of Policy Research*, *26*(4), 423–438. doi: 10.1111/j.1541-1338.2009.00393.x

Bush, G. W. (2001, October 29). *HSPD-1: Organization and Operation of the Homeland Security Council* (United States). Retrieved May 5, 2013, from http://www.fas.org/irp/offdocs/nspd/hspd-1.htm

Bush, G.W. (2001, October 29). *HSPD-2: Combating Terrorism through Immigration Policies* (United States). Retrieved May 5, 2013, from http://www.fas.org/irp/offdocs/nspd/hspd-2.htm

Bush, G. W. (2002, March 11). *Homeland Security Presidential Directive-3* (United States). Retrieved May 5, 2013, from http://www.fas.org/irp/offdocs/nspd/hspd-3.htm

Bush, G. W. (2002). *Securing the Homeland, Strengthening the Nation* (United States, Department of Homeland Security). Retrieved May 7, 2013, from http://georgewbush-whitehouse.archives.gov/homeland/homeland_security_book.html

Bush, G.W. (2002, September 17). *NSPD 17/HSPD 4 (unclassified version): National Strategy to Combat Weapons of Mass Destruction* (United States). Retrieved May 5, 2013, from http://www.fas.org/irp/offdocs/nspd/nspd-17.html

Bush, G. W. (2003, December 17). *Homeland Security Presidential Directive/HSPD-7* (United States). Retrieved May 15, 2013, from http://www.fas.org/irp/offdocs/nspd/hspd-7.html

Bush, G.W. (2003, February 28). *Homeland Security Presidential Directive-5* (United States). Retrieved May 5, 2013, from http://www.fas.org/irp/offdocs/nspd/hspd-5.html

Bush, G. W. (2003, September 16). *Homeland Security Presidential Directive/HSPD-6* (United States). Retrieved May 5, 2013, from http://www.fas.org/irp/offdocs/nspd/hspd-6.html

Bush, G. W. (2007). *National Strategy for Homeland Security* (United States, Office of Homeland Security). Washington, DC.

Cooper, C., and Block, R. (2006). *Disaster: Hurricane Katrina and the failure of Homeland Security*. New York: Times Books.

Dwyer, J., Flynn, K., & Fessenden, F. (2002, July 7). FATAL CONFUSION: A Troubled Emergency Response; 9/11 Exposed Deadly Flaws in Rescue Plan. *New York Times*. Retrieved May 2013 from http://www.nytimes.com/2002/07/07/nyregion/fatal-confusion-troubled-emergency-response-9-11-exposed-deadly-flaws-rescue.html?pagewanted=all

Hedgpeth, D. (2008, September 17). Congress says DHS oversaw $15 billion in failed contracts. *Washington Post*. Retrieved May 9, 2013, from http://articles.washingtonpost.com/2008-09-17/news/36802758_1_dhs-contracts-homeland-security

Homeland Security Act (2002), http://www.dhs.gov/xlibrary/assets/hr_5005_enr.pdf 2135 § 107–296.

Lanier, M. M., & Briggs, L. T. (2013). *Research methods in criminal justice and criminology: A mixed methods approach*. Oxford: Oxford University Press.

Larose, D. T. (2006). *Data mining methods and models*. Hoboken, NJ: Wiley-Interscience.

Lipton, E. (2006, July 19). Homeland Security Department is accused of credit card misuse. *New York Times*. Retrieved May 2013 from http://www.nytimes.com/2006/07/19/washington/19cards.html?_r=0

McAuliff, M. (2012, October 2). Homeland Security "fusion centers" target fishing, marriage counseling, at huge cost. *Huffington Post*. Retrieved May 2013 from http://www.huffingtonpost.com/2012/10/02/homeland-security-fusion-centers_n_1933998.html

Presidential Directives. (n.d.). *Homeland Security Digital Library (HSDL): Policy and Strategy Documents*. Retrieved May 22, 2013, from https://www.hsdl.org/?collection/stratpol&id=pd&pid=gwb

Reese, S. (2005). *Homeland Security Advisory System: Possible Issues for Congressional Oversight* (United States, Congressional Research Service, Library of Congress).

Schmitt, E., & Shanker, T. (2011). *Counterstrike: The untold story of America's secret campaign against al Qaeda.* New York: Times Books.

Sniffen, M. J. (2007, September 6). DHS ends criticized data-mining program [editorial]. *Washington Post.* Retrieved May 15, 2013, from http://www.washingtonpost.com/wp-dyn/content/article/2007/09/05/AR2007090500795.html

United States, Central Intelligence Agency. (n.d.). *Central Intelligence Agency.* Retrieved May 17, 2013, from https://www.cia.gov/about-cia/cia-vision-mission-values

United States, Department of Homeland Security. (2008). *National response framework.* Washington, DC: U.S. Dept. of Homeland Security.

United States, Department of Homeland Security. (2013). *Ready .gov.* Retrieved from http://www.ready.gov/about-us

United States, Department of Homeland Security, History Office. (2008). *Brief Documentary History of the Department of Homeland Security 2001–2008.*

United States, Department of Homeland Security. (n.d.). *About the National Cybersecurity & Communications Integration Center.* Retrieved May 7, 2013, from http://www.dhs.gov/about-national-cybersecurity-communications-integration-center

United States, Department of Homeland Security. (n.d.). *NSPD-59/HSPD-24 on Biometrics for Identification and Screening to Enhance National Security.* Retrieved May 17, 2013, from http://www.fas.org/irp/offdocs/nspd/nspd-59.html

United States, Department of Homeland Security. (n.d.). *Who Joined DHS.* Retrieved May 17, 2013, from http://www.dhs.gov/who-joined-dhs

United States, Federal Bureau of Investigation. (n.d.). *FBI.* Retrieved May 17, 2013, from http://www.fbi.gov/about-us/intelligence/mission

United States, Federal Emergency Management Agency. (n.d.). *About the Agency/FEMA .gov.* Retrieved May 15, 2013, from http://www.fema.gov/about-agency

United States, Government Accountability Office. (2004). *Homeland Security: Effective Regional Coordination Can Enhance Emergency Preparedness.* Retrieved May 23, 2013, from http://www.gao.gov/new.items/d041009.pdf

United States, Government Accountability Office. (2010). *U.S. GAO. Information Sharing: Federal Agencies Are Helping Fusion Centers Build and Sustain Capabilities and Protect Privacy, but Could Better Measure Results.* Retrieved May 2013 from http://www.gao.gov/products/GAO-10-972

United States, Government Accountability Office. (2011). *DATA MINING: DHS Needs to Improve Executive Oversight of Systems Supporting Counterterrorism.*

United States, Government Accountability Office. (n.d.). *GAO-06-957T, Purchase Cards: Control Weaknesses Leave DHS Highly Vulnerable to Fraudulent, Improper, and Abusive Activity.* Retrieved May 22, 2013, from http://www.gao.gov/htext/d06957t.html

United States, Homeland Security. (2008). *National Incident Management System.* Retrieved May 16, 2013, from http://www.fema.gov/pdf/emergency/nims/NIMS_core.pdf

United States, Secret Service. (n.d.). *United States Secret Service: Vision and Mission Statements.* Retrieved May 17, 2013, from http://www.secretservice.gov/mission.shtml

Waugh, W. L. Jr., & Streib, G. (2006). Collaboration and leadership for effective emergency management. *Public Administration Review.*

Wise, C. R., & Nader, R. (2002). Organizing the federal system for Homeland Security: Problems, issues, and dilemmas. *Public Administration Review*, 62(S1), 44–57. doi: 10.1111/1540-6210.62.s1.8

Witten, I. H., and Frank, E. (2005). *Data mining: Practical machine learning tools and techniques.* Amsterdam: Morgan Kaufman.

Appendix A

First Seven Homeland Security Presidential Directives

Table A1:Description of the First Seven Homeland Security Presidential Directives

Presidential Directive Number	Year	Name of Directive	Description of Directive
HSPD-1	2001	Organization and Operation of the Homeland Security Council	Ensures coordination of all homeland security-related activities among executive departments and agencies and promote the effective development and implementation of all homeland security policies.
HSPD-2	2001	Combating Terrorism Through Immigration Policies	Provides for the creation of a task force which will work aggressively to prevent aliens who engage in or support terrorist activity from entering the United States and to detain, prosecute, or deport any such aliens who are within the United States and to detain,prosecute, or deport any such.
HSPD-3	2002	Homeland Security Advisory System	Establishes a comprehensive and effective means to disseminate information regarding the risk of terrorist acts to Federal, State, and local authorities and to the American people.
HSPD-4	2002	National Strategy to Combat Weapons of Mass Destruction	Applies new technologies, increased emphasis on intelligence collection and analysis, strengthens alliance relationships, and establishes new partnerships with former adversaries to counter this threat in all of its dimensions.
HSPD-5	2003	Management of Domestic Incidents	Enhances the ability of the United States to manage domestic incidents by establishing a single, comprehensive national incident management system (NIMS).
HSPD-6	2003	Integration and Use of Screening Information	Provides for the establishment of the Terrorist Threat Integration Center.
HSPD-7	2003	Critical Infrastructure Identification, Prioritization, and Protection	Establishes a national policy for Federal departments and agencies to identify and prioritize United States critical infrastructure and key resources and to protect them from terrorist attacks.

Appendix B

Governmental Structure Before and After DHS

Many government agencies with important roles in the Department of Homeland Security were once in other departments. These agencies were moved in order to achieve the goal of reinforcing security and streamlining communication.

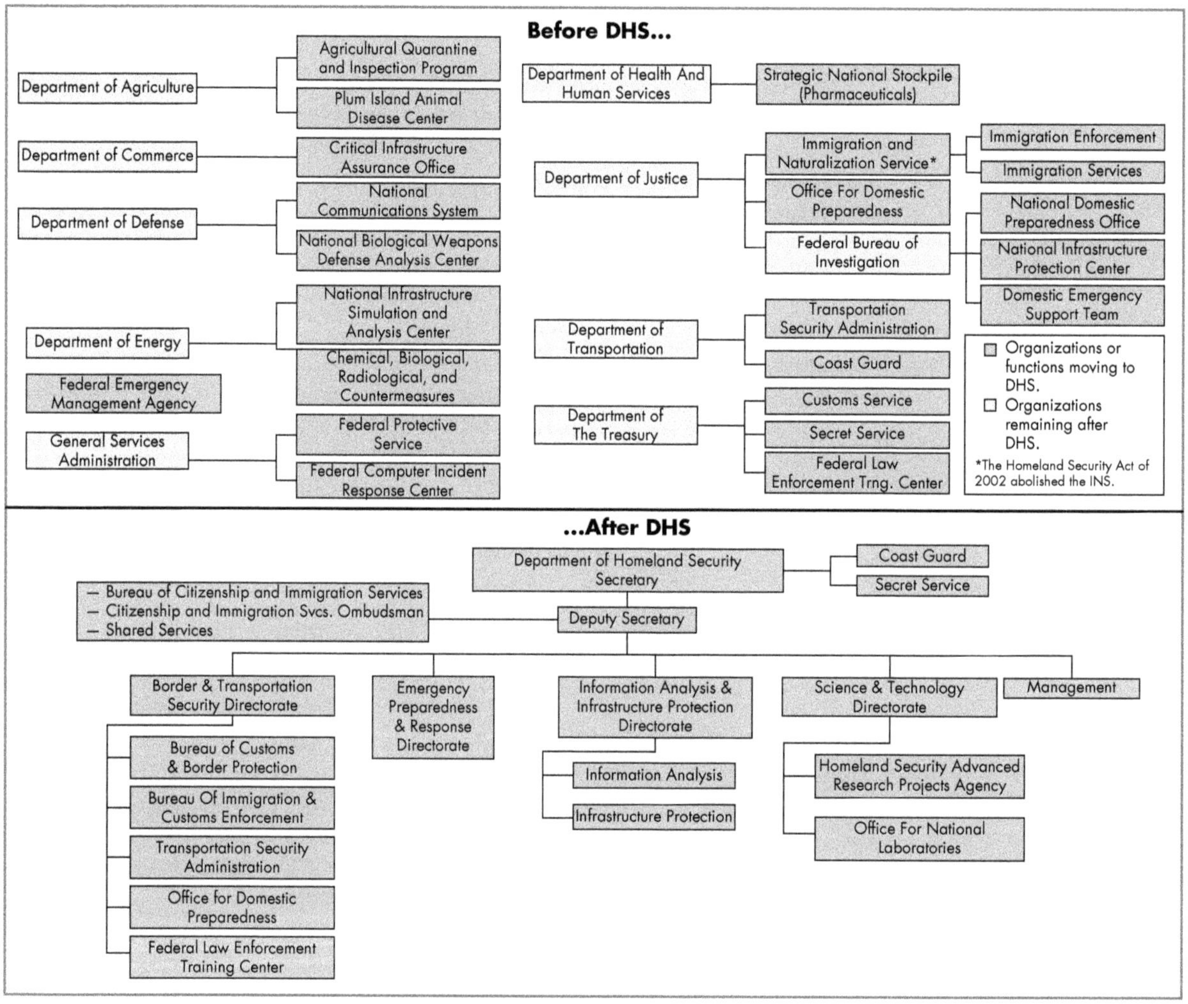

Figure B1

Appendix C

Hierarchical Structure of DHS

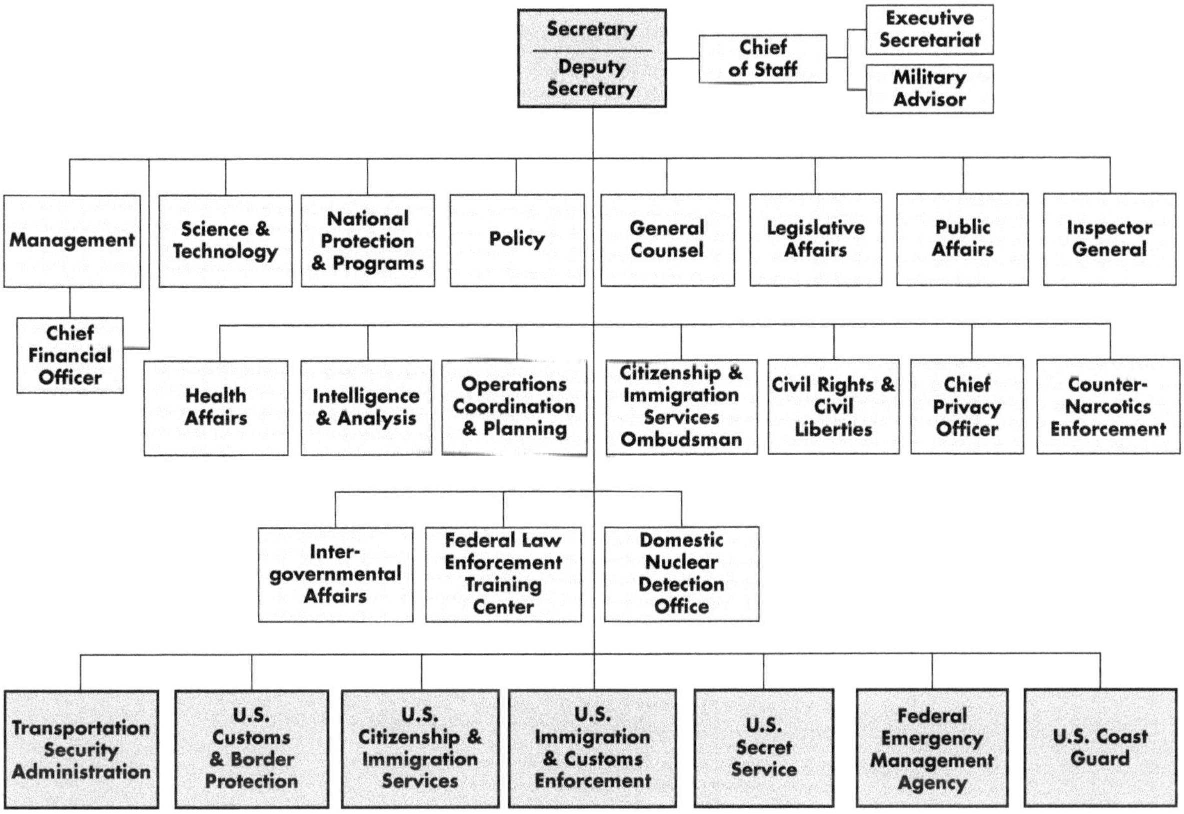

Figure C2

Appendix D

Communication Networking Under DHS

One of the main goals of the Department of Homeland Security is to create effective lines of communication among a variety of agencies, programs, and services. Most of these government entities are under transportation management centers, emergency operations centers, and fusion centers.

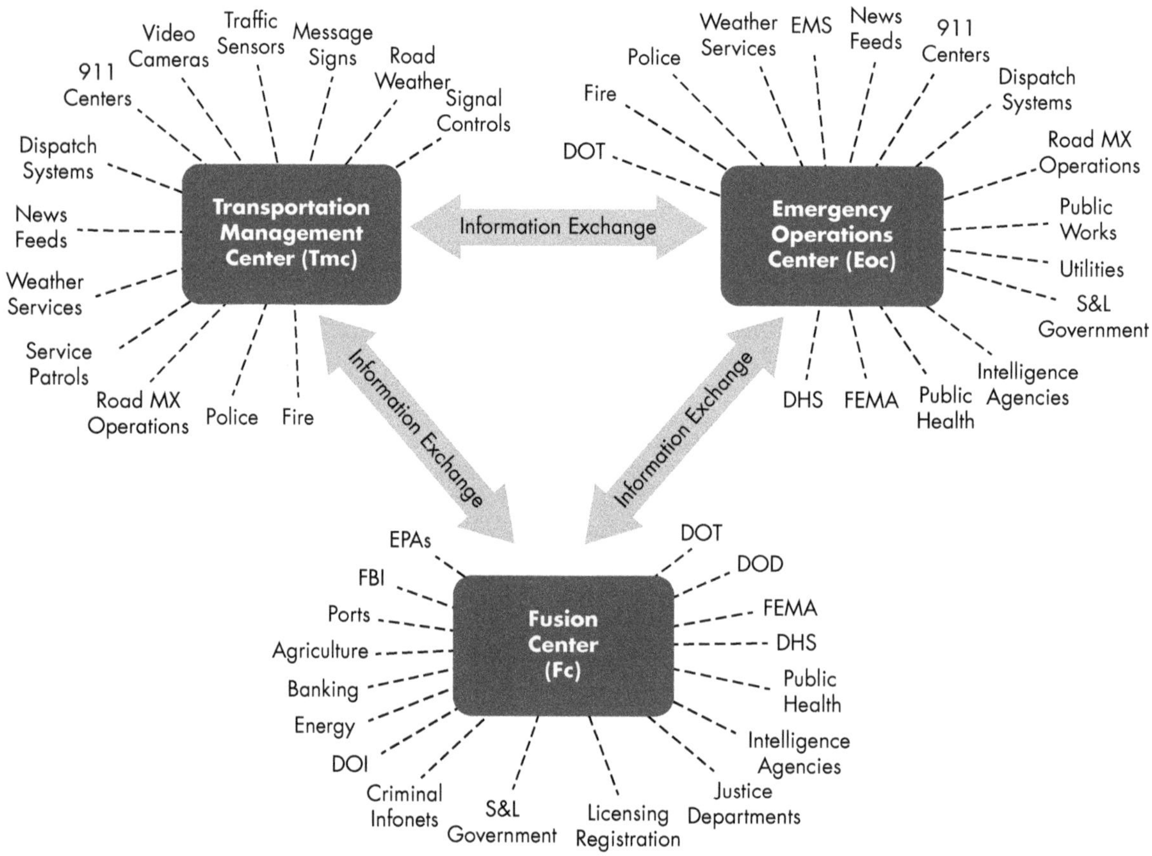

Figure D3

Appendix E

Cyber Security Chain of Command

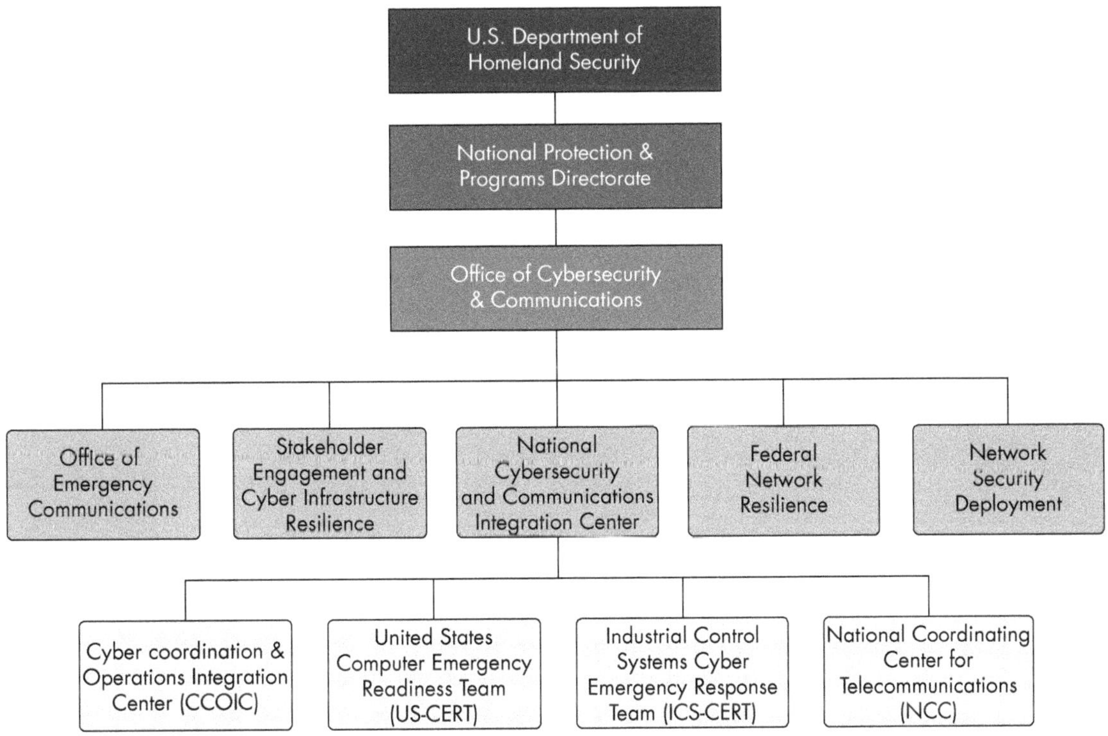

Figure E4

Chapter 2

Criminal Justice and Terrorism: The Era of Homeland Security

By Herbert A. Johnson, Nancy Travis Wolfe, and Mark Jones

The Path to 9/11: A Brief History of Terrorism

One of the most worn-out clichés in recent years is, "The events of September 11, 2001 changed America forever." Even though it is a cliché, it is true. Anyone old enough to remember can recall where they were and what they were doing when they heard the news of the September 11 terrorist attacks on New York City and Washington, D.C. Many people in the United States separate their lives into the pre- and post-9/11 eras, just as the World War II generation did with the December 7, 1941, attack on Pearl Harbor. Many aspects of American foreign and domestic policy are divided the same way. Because September 11, 2001, was the catalyst for so much change, those who remember it and are living in its aftermath would do well to understand the events that led up to it.

The roots of modern Islamic terrorism are complex and tied to issues other than the ones now associated with such terrorism, namely extremist Islam, sometimes called Islamism. Arab nationalism, European colonialism, pride in Islamic history, and other factors have contributed to the current state of Islamic terrorism which to many Americans, seem to have come from nowhere. The events of 9/11 were (in part) the culmination of several centuries of events that took place both inside and outside the Arab and Muslim worlds.

Understanding the Middle East, let alone dealing with its problems, has never been simple. The Middle East has been a source of frustration not only for its own leaders, but for those in Europe and the United States as well. Fervent religious beliefs, based in part because three of the world's major religions—Judaism, Christianity, and Islam—had their beginnings in the Middle East, combined with racial, ethnic, and tribal

Terrorism and Homeland Security

1790	Edmund Burke, British political philosopher, describes the French Revolution as a "reign of terror."
1807–1814	During the Peninsula War, the term *"guerilleros"* (irregulars) is used to describe the tactics of asymmetrical warriors.
1901	President William McKinley is assassinated in New York by an anarchist terrorist.
1910	The office of the *Los Angeles Times* is blown up, killing 21.
1914	The heir to the throne of the Austro-Hungarian empire is assassinated by a Bosnian Serb terrorist in Sarajevo, igniting World War I.
1917	The Balfour Declaration declares support for a Jewish state in the land of Palestine.
1918	The end of World War I signals the end of Ottoman rule of the Middle East, and countries' boundaries are carved out by European colonial interests.
1919	Michael Collins assumes control of the militant wing of the Irish Republican Army and organizes attacks against occupying British forces in Ireland
1921	Ireland gains independence from Britain, but Northern Ireland remains part of the United Kingdom, propelling future terrorist acts by Protestant and Catholic militants.
1928	The Muslim Brotherhood is founded in Egypt.
1946	Members of the Irgun, a Zionist terrorist organization, blow up the King David Hotel in British-occupied Palestine, killing 91.
1946–1991	The Cold War pits American interests versus those of the U.S.S.R., resulting in the covert funding of terrorist activities allied with both nations' interests.
1948	The nation of Israel is established, the culmination of 70 years of gradual Jewish migration to Palestine.
1948	Egyptian Prime Minister is assassinated by the Muslim Brotherhood.
1949	Hassan Al-Banna, the founder of the Muslim Brotherhood, is killed by Egyptian government.
1964	The Palestine Liberation Organization (PLO) is founded.
1969	Yassir Arafat assumes control of the PLO.
1972	Israeli athletes are murdered by Palestinian terrorists at the summer Olympics in Munich.
1979	A revolution in Iran installs a radical Shiite government, and results in the storming of the American embassy in Tehran, with American personnel being held hostage for 444 days.
1979	The Soviet Union invades Afghanistan, triggering an American supported Muslim Jihad against the Soviets.
1981	As retaliation for his peace overtures toward Israel, Egyptian President Anwar Sadat is assassinated by soldiers allied with the Muslim Brotherhood and Egyptian Al-Jihad.
1982	Israel invades Lebanon, which results in the creation of Hezbollah.
1983	241 American Marines are killed by a Hezbollah-led suicide bomber in Beirut, Lebanon.
1988	Pan Am Flight 103 explodes over Lockerbie, Scotland, killing all passengers and some on the ground; a Libyan agent is eventually convicted of the bombing.
1989	The Soviet Union withdraws from Afghanistan, paving the way for the takeover by the Taliban and the establishment of Al Qaeda, led by Osama bin Laden.
1993	Terrorists bomb the World Trade Center in New York City, killing six.
1995	A bomb destroys the Murrah Federal Building in Oklahoma City, killing over 200; two right wing American extremists are convicted of the bombing.

1996	Terrorists bomb the Khobar Towers in Saudi Arabia, killing 19 Americans and one Saudi.
1998	Osama bin Laden declares war on the United States, and names any American as a potential target.
1998	Al Qaeda terrorists bomb the American embassies in Kenya and Tanzania, killing over 200.
2000	Al Qaeda suicide bombers attack the *USS Cole* in Yemen, killing 17 sailors.
2001	Al Qaeda terrorists crash four commercial airliners into both towers of the World Trade Center, the Pentagon, and in a field in Shanksville, PA, killing over 2,900.
2001	The Office of Homeland Security is created.
2001	The United States invades Afghanistan, deposes the ruling Taliban and scatters Al Qaeda.
2003	An American-led coalition invades Iraq and deposes Saddam Hussein
2004	Al Qaeda-inspired terrorists blow up a train in Madrid, Spain in retaliation for Spanish involvement in Iraq.
2005	Al Qaeda-inspired terrorists attack London's transportation system, killing 52.

conflicts, not to mention the fact that the Middle East is a source of important natural resources and is a primary transit point connecting all points of the world, have added to this complexity.

Part of the problem when dealing with the Middle East is that almost every ethnic or religious group residing there aspires to a past glory, or believe they have a historical or divine claim to disputed territory. Modern Israelis believe that the nation of Israel should exist either by divine right or historical inheritance, and that the world's Jews need a homeland they can call their own. Those who claim divine right point to the glory of the Old Testament kingdoms of David and Solomon, and to the entire Old Testament which depicts the Jews as God's chosen people.

Christians in the Middle East, as well as many Christians throughout the world, see the Middle East, especially Israel, as a sacred site as well. Modern-day Israel was the place where Jesus lived. Apocalyptic (end of world) prophecies center on the Middle East as the place where the world will come to an end.

The Arabian Peninsula was the birthplace of Islam, and Saudi Arabia is home to Islam's holiest cities, Mecca and Medina. It was from modern-day Saudi Arabia that the prophet Mohammed and the early disciples of Islam, most of whom were Arabs, set about spreading Islam to the rest of the world. The seventh through the sixteenth centuries are often referred to as the Golden Age of Islam because they witnessed the spread of the Islamic faith. During the Golden Age of Islam, the Muslim religion would spread to non-Arabs as far east as Indonesia, as far south as eastern Africa, as far west as Northern Africa, and as far north as Bosnia and Spain. Not only do many Muslims long for a return to the Golden Age of Islam, many believe that the Middle East is theirs by divine right.

Another Islamic holy city is Jerusalem, which is also a holy city to Jews and Christians. During the twelfth and thirteenth centuries, European armies, with the blessing of the Roman Catholic hierarchy, sought to reclaim the Middle East for Christianity and to rid Christianity's holy land of the "infidel" Muslims who had dominated it for several centuries. The Crusades rank among the bloodiest and most bitter conflicts in human history. To this day, some Muslims, Osama bin Ladin among them, derisively refer to all Westerners who seek influence in the Middle East as "crusaders," realizing that the term still resonates with Middle Eastern Arabs and Muslims who have learned of the Crusades by studying Islamic history.

Terrorism's Origins

To some degree, terrorism is an offshoot of guerrilla warfare. The term "guerrilla" is rooted in the early nineteenth century with the French occupation of the Iberian Peninsula. The occupying French forces were far superior to the Portuguese and Spanish insurgents, known as *guerrilleros* or "irregulars," so the Spanish and the Portuguese *guerrilleros* employed hit-and-run tactics and made a practice of keeping the French off-guard.[1] This type of asymmetrical warfare has been employed very often when a smaller or less well-equipped force encounters superior conventional forces. American patriots used it against the far superior British forces during the American Revolution; it was used against Americans during the Vietnam conflict.

Whether a person or organization should wear the label of terrorist is fiercely debated. There are many who bristle at the idea of comparing modern terrorists with any sort of soldier, be it guerrilla or any other kind. While a person who engages in acts of violence or killing may be called a terrorist by one person, that same individual or organization may be called a freedom or resistance fighter by someone else, regardless of how cruel their actions may seem. Like the typical guerrilla warrior, a terrorist may see him- or herself as having to resort to unconventional means of warfare such as targeting civilians, because they cannot win through conventional warfare. A group of 19 individuals like the September 11, 2001, hijackers could not possibly hope to compete directly with the United States military in a conventional battle, but that same group of people could change the course of history if they are willing to commit an act as dastardly as crashing a commercial civilian airliner into a building filled with people who are completely defenseless. No matter how one views the events of September 11, 2001, one cannot deny that it was a date that changed history. Tom Friedman of the *New York Times* states that it was nothing less than the beginning of World War III, which is still unfolding.[2] By no means did terrorism originate on September 11, and by no means has it ever been the sole province of Muslim militants. Many other causes, ideologies, and personality types have been at the root of terrorist behavior for more than 100 years.

Anarchism and Labor Militancy

In the eyes of many Americans, the image of the modern terrorist is an angry young Muslim, but this was not always the case. First, it should be noted that most of the world's Muslims do not endorse terrorism, let alone participate in terrorist activities, and it is unfortunate that people of Middle Eastern or South Asian extraction live with this stereotype. However, at the beginning of the twentieth century it was not Muslims who wore the face of the terrorist. Instead, the image of the terrorist was a (sometimes) hooded or masked Irishman or Eastern or Southern European. The ideology connected to these terrorists was not Islam, but socialism, communism, anarchism, and/or a strong allegiance to organized labor.

The dawn of a new century always brings with it the hope that it will be more peaceful than the last, but this hope seems never to become reality. On September 6, 1901, a notorious act of terrorism occurred at the Pan-American Exposition in Buffalo, New York. Leon Czolgosz, a self-described anarchist, shot and mortally wounded President William McKinley, who died eight days after the shooting. Although Czolgosz was not mentally coherent enough to serve in a leadership capacity in any organization, the image of the anarchist terrorist as madman, already at the forefront of American images of terrorism, was greatly reinforced.[3] This image was reinforced when Emma Goldman, a leader in the anarchist movement who had met Czolgosz but did not participate in the McKinley murder, praised Czolgosz for following the dictates of his ideals. Buffalo authorities treated Czolgosz roughly and he was beaten repeatedly and nearly lynched by Buffalo crowds. Despite his apparent mental infirmity, Czolgosz was tried and convicted of the President's murder. Czolgosz was executed at Auburn Prison on October 20, 1901, less than two months after McKinley's death.[4]

Terrorism and Labor Militants

The McKinley assassination was only the beginning of left-wing radical terrorism in the twentieth century. In many respects an offshoot of communism and anarchism, labor unions, while now mainstream, were viewed by opponents as subversive to American capitalism while it was in its infancy. One of the worst mass killings at the hands of terrorists in American history occurred on October 1, 1910, in Los Angeles at the hands of organized labor militants. Early that morning, a dynamite explosion at the *Los Angeles Times* headquarters resulted in 21 deaths and many more injuries. At the time, the *Los Angeles Times* was a vociferous opponent of organized labor.

Local police departments, even those in large cities like Los Angeles, lacked the expertise to investigate complex crimes as well as the resources to conduct interstate investigations. The bomb was an instrument of war, not of crime, and police agencies were not qualified to conduct a proper investigation of such a sophisticated weapon. The only federal law enforcement agency that could have become involved was the recently created Bureau of Investigation, but it was hampered by poorly qualified agents and few resources. In addition, in marked contrast to the modern era, the federal government was not as likely to get involved in what was viewed as a state or local matter; so the administration of President William Taft did not feel obligated and was not pressured to offer any appreciable assistance to the Los Angeles investigation. Crime, even on this scale, was simply a local matter and was not viewed as being within the purview of the federal government.

George Alexander, the Mayor of Los Angeles, contacted William Burns, head of the Burns Detective Agency, who was already investigating a similar crime that had occurred at a railroad office in Illinois. Burns was awarded a contract from the local government to fund the investigation. Burns' investigation began with much promise but met some political roadblocks; namely, his funding was cut off as a result of a dispute with the Los Angeles District Attorney. Despite having to use his own funds (He eventually received a $200,000 reward for solving the case.), Burns investigated the bombing. Three labor militants, Ortie McManigal, James McNamara, and John McNamara, all of whom later told Burns they would blow up the whole country if it meant obtaining worker rights, were charged with the bombing.

The American Federation of Labor hired famed criminal defense attorney Clarence Darrow to defend the trio. Despite impassioned and nationally publicized accusations of a capitalist conspiracy against organized labor, the case went to trial. While the trial was being conducted, the three defendants changed their plea to guilty and were sentenced to long prison terms. The trial was rife with accusations of tampering on both sides. Burns was indicted for kidnapping in Indiana, where he had gone to apprehend the bombers to forcibly transport them to California. He was not convicted. Darrow was eventually charged with attempting to bribe the jurors but was acquitted as well.[5]

World War I

One of the most notorious acts of political assassination/terrorism occurred on June 28, 1914. Bosnia, which then (as now) contained a large population of ethnic Serbians, was under the dominion of the Austro-Hungarian Empire. Many Bosnian Serbs were seeking independence from the Austro-Hungarian Empire and wanted a separate Serbian state within Bosnia—the same issue that provoked the mass killings in Bosnia in the 1990s. Among those seeking this independence was a group called the Serbian Black Hand, and a group they supported called Mlada Bosna, or Young Bosnia, which counted among its members 19-year-old Gavrilo Princip. Despite prior warnings by the Serbian ambassador to Austria of a potential assassination attempt, Archduke Franz Ferdinand, the Crown Prince and heir to the Austrian throne, accepted an invitation from the Bosnian governor to visit Sarajevo, Bosnia, and inspect his army troops. While riding through Bosnia, another member of Mlada Bosna named Nedelijko Cabrinovic threw a grenade at the Archduke's car. The Archduke survived, but rather than getting out of Sarajevo as quickly as possible, he decided to visit the local hospital to visit those wounded in the assassination attempt. After the visit, while riding through Sarajevo, his driver took a wrong turn and happened to drive by Princip, who had given up his planned attempt to kill the Archduke when the grenade attack went awry. Princip took advantage of the Archduke's fortuitous reappearance, and shot and killed the Archduke and his wife.[6]

Although the assassination generated big headlines in much of Europe, it went relatively unnoticed in the United States. However, over the next few weeks, a series of threats were exchanged, ending with a declaration of war by Austria-Hungary and Germany (an Austrian ally) against Serbia, a traditional Russian ally. Before long, other countries aligned with one side or the other joined the fray, and World War I, also referred to later as The Great War or The War to End All Wars, began. Three years later the United States entered the war on the side of Serbia and its allies, which included Great Britain and France, as well as Italy and Japan, two countries that the United States would square off against 24 years later in World War II. Princip, a low-level soldier in a ragtag group of young radical terrorists, pulled the trigger that resulted in the deaths of approximately 20 million people and a reordering of the entire world order.

Part of the new order brought about by World War I's end was the breakup of the Ottoman Empire. Once the most powerful empire in the Middle East and Southern Europe, the Ottoman Empire had been in gradual decline prior to World War I. The Ottomans lost most of their European possessions in the First Balkan War of 1912–1913. World War I brought the official end to the Ottoman Empire, and dominion over its territories was divided among Russia, Great Britain, France, and Italy. The Middle East was divided to suit the economic and political interests of the victorious Europeans, with little regard for the ethnic and tribal divisions within those countries. To this day, many Middle Easterners do not define their identity with the nation in which they live. They identify themselves with their ethnic, tribal or religious affiliation more than they do nationality, to the delight of those fighting against Middle Eastern governments.

The creation of these somewhat artificial nation-states is still a bone of contention throughout the Middle East, as was an announcement by Arthur Balfour, the British foreign minister, with respect to Palestine. The "Balfour Declaration" as it came to be called, expressed sympathy for the idea of creating a sovereign Jewish state in Palestine. For several decades Palestine had witnessed a gradual influx of Jewish immigration, along with calls for the creation of a Jewish state. The Balfour Declaration also called for equal treatment of non-Jews in Palestine, namely Palestinian Arabs.[7] Over the next three decades, Jewish immigration to Palestine would increase and so would Jewish influence in Palestine, and the calls for the creation of a Jewish state would become louder, especially as persecution of Jews throughout Continental Europe escalated in the 1930s.

Michael Collins and the Irish Republican Army

The seeds of modern terrorism did not take root exclusively in the Middle East or the European Balkan countries like Serbia and Bosnia. While the Middle East was undergoing its forced political evolution, so was another traditionally British-dominated country, Ireland. Given their revulsion of the West, its politics, culture, and religion, it would seem unlikely that modern Islamic terrorists would draw inspiration or learn lessons from an Irish Catholic. Yet Michael Collins, whether knowingly or not, stands as one of the important figures in twentieth-century terrorism. Born on October 16, 1890 in Clonakilty, Ireland, Collins grew up amid rather unremarkable circumstances. There were few indications in his early life that Collins would be viewed alternately as a merciless terrorist and as the voice of Irish patriotism in the early twentieth century. Collins worked rather anonymously in the British civil service from 1906–1916, although he joined Irish nationalist groups in 1908 and 1909. Ireland had lived under complete British domination since 1800, and the two countries had endured extremely poor relations for hundreds of years.

Collins participated in the Easter Rising in 1916—a disastrous (for the rebels) uprising against the British at the height of World War I. He was imprisoned for his role in the uprising and released in December 1916. Like many terrorists or resistance fighters, the Easter Rising convinced Collins that the opposition, which in this case was the British, could not be defeated or forced to negotiate Irish independence by either diplomacy or conventional warfare. Collins was elected to the Irish Assembly in 1918, representing Sinn Fein (Ourselves Alone), the political wing of the Irish Volunteers, which was the precursor to the Irish Republican Army.

Michael Collins assumed command of the Irish Republican Army in 1919 and orchestrated attacks against the British, employing guerrilla tactics, bombings, and assassinations of British officials, intelligence agents, soldiers, and police officers as a way of forcing the British to the negotiating table. For a time, Collins was at the top of Britain's most wanted list, with a $10,000 bounty on his head. Collins studied the methods used by Russian revolutionaries in the late nineteenth century.[8] Based on his studies of the Russians, Collins thought that bombs, ambushes, and assassinations of British soldiers and officials and their Irish supporters would be successful weapons in Ireland's quest for independence.

Whether it was a result of the Collins-led IRA terror campaign or international pressure, the British sat down at the negotiating table with their archenemy in the fall of 1921. Collins and other IRA negotiators persuaded the British to grant independence to most of Ireland, but the country's Ulster region, which consisted of six counties and a sizeable Protestant population, would remain under British control and as part of the United Kingdom, as it does even today. While the Treaty was greeted enthusiastically by many of the Irish, a substantial minority, including most of the IRA leadership, were outraged by the Northern Ireland compromise and called Collins a traitor.

Collins was killed on August 22, 1922 at the age of 31. He was killed in an ambush by opponents of his efforts, although Collins biographer Peter Hart suggests that Collins was not intentionally targeted. Since his death, Collins has been the subject of admiration, fascination, and revulsion. He embodies the debate about whether one person's terrorist is another person's freedom fighter. His life and methods of operating, which he gleaned by studying Russian revolutionaries, were studied and copied by left-wing and Muslim extremists alike. Although he has been dead for more than 85 years, and he could not have foreseen the impact his life would have on terrorist and nationalist movements beyond his native Ireland, Collins simultaneously wears two faces—the embodiment of Irish nationalism and the epitome of modern terrorism.[9]

The fact that not all of Ireland was unified as a result of the 1921 treaty meant that the IRA continued to exist, even though it was officially outlawed by the new Republic of Ireland, and it continued to stand as one of Great Britain's worst enemies. For the rest of the twentieth century, the IRA was a major thorn in the side of the British, and untold numbers of lives were lost on both sides as a result of IRA bombings and assassinations, as well as British reprisals and crackdowns against the IRA. In addition, many killings in Northern Ireland are attributable to radical Protestant groups who favor the union with the United Kingdom and oppose the IRA. Currently both the United Kingdom and the United States State Department lists remaining branches of the IRA as terrorist organizations.

Listed below are some of the most notorious acts of terrorism attributed to the IRA over the past four decades:

- 1972—At the Aldershot headquarters of the British Parachute Regiment, an IRA bomb killed seven, none of them British soldiers
- 1973—IRA car bombs in London killed one and injured 180
- 1974—An IRA bomb killed 11 British soldiers and family members
- 1974—An IRA bomb in Guildford killed two soldiers and three civilians and injured 50 others
- 1974—An IRA bomb at a pub in Birmingham killed 21 (This bombing formed the basis for the movie *In the Name of the Father.*)
- 1979—Louis Mountbatten, a British admiral, statesman, and uncle to Prince Philip, was assassinated by the IRA in Ireland
- 1984—At the British Conservative Party gathering in Brighton, an IRA bomb killed five, including a prominent Conservative party leader
- 1993—In Warrington, an IRA bomb killed two boys, ages 3 and 12, while they were shopping with their families[10]

While the IRA was waging its campaigns, and it is not completely extinct even today, other terrorist groups with other agendas waged campaigns of their own. The IRA formed unlikely partnerships with some of these groups despite their significant religious and ethnic differences. Throughout the twentieth century the IRA and radical Arab governments shared resources and expertise in training their members.

The Muslim Brotherhood

One of the seminal events in the history of modern Islamic terrorism is the establishment of the Muslim Brotherhood. In fact, most Muslim extremism and terrorism can be traced to the establishment of the Muslim Brotherhood in Egypt. The Muslim Brotherhood was founded in 1928 by Hasan al-Banna, a 22-year-old school teacher. The Brotherhood and its radical Islamic idealism quickly spread to North Africa and the Middle East.

At first the Brotherhood was primarily a religious organization with little involvement in politics; they were primarily a social activist organization rather than a political movement, even though many of their members were active opponents of European colonial influence in their countries. The Brotherhood advocated a lifestyle of strict adherence to Quranic principles. Al-Banna and the Brotherhood wanted to reduce or eliminate the influence of Westerners, Christians, and non-Muslims in the Islamic world, especially in the Arab domains of the Middle East and North Africa.

During the late 1930's, the Brotherhood began to assume a more active posture on political matters, advocating not only a Muslim lifestyle, but a theocratic government based on Islamic principles as well; in short, a political Islam. Like many other Egyptians, the Brotherhood wanted to end the British occupation of Egypt, and they had little use for the ruling monarch, King Farouk. In December 1948, a member of the Brotherhood assassinated Egypt's Prime Minister, Mahmoud an-Nukrashi Pasha, after the government launched a crackdown on the Brotherhood. Hasan Al-Banna, the founder of the Muslim Brotherhood, was assassinated in retaliation in February 1949.

The Brotherhood survived al-Banna's death, but it quickly found an enemy in the secular government headed by Gamal Abdel Nasser. Although Nasser was an observant Muslim in his personal life and, like the Brotherhood, a vehement opponent of the nation of Israel, he detested the idea of political Islam. The Brotherhood attempted to assassinate Nasser in 1954, which led to a severe crackdown on the Brotherhood. During the 1950s and the 1960s, the Brotherhood's influence was eclipsed by Arab and Muslim leaders who adopted an ideology often referred to as "Nasserism," or Arab Socialism.[11]

The Brotherhood experienced a resurgence in the 1970s and 1980s, spurred on by the Iranian revolution of 1979 and disillusionment with secular Nasserism. Although several thousand Brotherhood members were killed in a 1982 uprising by President Hafez Assad's government in Syria, the Brotherhood's influence widened. Even though it is still officially outlawed in Egypt, the Brotherhood has representatives in both the Egyptian and Jordanian governments.[12]

The establishment of the Brotherhood stands as an important event in the history of twentieth century terrorism because it was a pioneer in the use of violence to impose a strict, some would say perverted, version of Islamic government in Islamic countries. Some of their adherents have been instrumental in trying to spread this ideology to non-Muslim areas as well, including Western Europe and North America.

Zionist Terrorism

There are some places in Israel (and the disputed territories in the West Bank, an area referred to by most Israelis as Judea and Samaria) even today where Arab schoolchildren must sometimes be escorted to school under armed guard. The reason is that some Jewish extremists want all Arabs off what they see as Jewish land. The Arabs are not alone in the use of violence and terrorism in Israel to further their agenda. In fact, the only Prime Minister in Israeli history to be assassinated was Yitzhak Rabin, who was killed on November 4, 1995, not by an Arab but by a right-wing Jewish extremist who opposed Rabin's peace overtures toward Yasser Arafat and the Palestine Liberation Organization. The assassin had links to EYAL (Fighting Jewish Organization), an outlawed and now-defunct Israeli-based terrorist group. On February 25, 1994, a Jewish terrorist who belonged to the (subsequently outlawed) Kach organization, which was an offshoot of the U.S.-based Jewish Defense League, killed 29 praying Muslims in Hebron before the surviving worshippers killed him. In addition to Arab terrorist groups, the Israeli government also concentrates attention on a few Jewish extremist groups as well, including Kahane Chai, an offshoot of the Kach and Jewish

Defense League. Kahane Chai calls for the total elimination of Arabs from all Jewish lands and the establishment of a Jewish the ocracy akin to the Old Testament kingdoms of David and Solomon.[13]

Terrorism and political assassination in the name of Jewish nationalism is not new. During the New Testament era when Jewish Palestine was being ruled by the Romans, a Jewish group called the Sicarii (dagger) instilled terror in Roman authorities and Jewish collaborators by carrying out assassinations in crowded places, often in broad daylight, and then disappearing into the crowd. This same group initiated a disastrous revolt against the Romans around 65 A.D. that led to the virtual destruction of the city of Jerusalem and the Jewish Diaspora, when the Jews were permanently scattered.[14]

From the time that the Zionist movement to resettle Israel as a Jewish homeland began in earnest in the late nineteenth century, some Zionists used violence to further their agenda. The principal Zionist terrorist organization was the Irgun Zvai Le'umi (National Military Organization). The Irgun's principal targets were Arabs and the British soldiers and officials who occupied Palestine during the British Mandate period of the 1930s and 1940s.

During the struggle for Israeli independence in the 1930s and 1940s, opinions of the Irgun among Jewish Zionists were split. Many Zionists favored diplomacy over force and despised the Irgun; the Irgun responded that most modern nations, including the United States, could not have been established without the use of force, and that no ethnic group could compare to the Jews when it came to persecution. The Irgun claimed that force was the only mechanism that would bring the British to the bargaining table for independence. They also claimed that some of their attacks were in retaliation for similar acts of violence perpetrated against Jews by Arab terrorists.

The Irgun leadership included men who some modern Israelis regard as heroes; chief among them was Menachem Begin, the last leader of the Irgun prior to the establishment of the state of Israel in 1948. During his time as head of the Irgun, from 1943–1948, Begin ordered and participated in a number of acts of violence, the worst of which was the bombing of the King David Hotel on July 22, 1946. The King David Hotel served as headquarters for British government, military, and police officials. However, not all of the deaths were military or government officials. The bombing killed 91, including many civilian employees at the Hotel. The Irgun claimed they provided advance warning of the bombing, but this did little to appease the feelings of those who were outraged by the bombing. Menachem Begin ordered the bombing, and a bounty was placed on his head by the British authorities.[15]

In a prototypical example of a terrorist-turned-statesman, Begin, a Polish-born Jewish immigrant to Palestine, eventually became one of the most prominent statesmen in Israeli history. He served as Prime Minister from 1977–1983, the first Israeli Prime Minister from the traditionally hardline Likud Party. Given his ultra-Zionist and Likud background, it is ironic that Begin, one of the most fervent Israeli nationalists in history, shared the 1978 Nobel Peace Prize with Egyptian President Anwar Sadat, his former archenemy, as a result of the Camp David Peace Treaty brokered by U.S. President Jimmy Carter. Even many years since his 1992 death, Begin's legacy is mixed among Israelis and non-Israelis alike.

The Establishment of Israel

For several decades prior to World War II, there was a growing movement to establish a stronger Jewish presence in Palestine, or the Jewish ancestral land of Israel. From the 1880s through World War II in the 1940s, the number of Jewish immigrates to Palestine increased gradually.

The European Holocaust of the 1930s and 1940s, which cost millions of Jews (and other ethnic groups) their lives, led many Jews and non-Jews alike throughout the world to believe that a Jewish homeland was necessary. During and after World War II, many Jewish refugees found themselves without a nation and with nowhere to go. In addition, the financial and political devastation of World War II signaled the beginning of the end of European colonial domination of Africa and the Middle East. The time was ripe for Zionists to seize control of Palestine and establish a Jewish state.

Using a combination of diplomacy and violence (see Time Capsule: Zionist Terrorism) the Jews in Palestine and their allies around the world brought the British, who had been occupying Palestine under a mandate, to the negotiating table, and the issue of Jewish statehood came before the United Nations. The United Nations voted to partition Palestine into two independent states, one for the Jews and one for the Arabs, but only one independent state emerged, and the nation of Israel was established in 1948. Ever since its inception, Israel and the countries that support it, including the United States, have been the target of terrorist attacks of all types, including Muslim militancy, left-wing terrorists, and right-wing terrorists.

Terrorism, the Cold War, and Arab Nationalism

The Cold War that pitted Western capitalism against Soviet Communism played an important role in the picture of terrorism from the end of World War II until the dissolution of the Soviet Union in the early 1990s. After World War II, much of the world, including almost all of Europe, parts of Southeast Asia, the Korean Peninsula, parts of Africa and the Middle East, and some countries in the Western Hemisphere were essentially divided into two camps, those who supported Western-style capitalism and democracy, and those who supported Soviet-style Communism or one of its variants. While the United States and the Soviet Union never confronted each other in battle during the Cold War, many countries served as locations for proxy wars between the two superpowers. In the cases of Korea, Vietnam, and Afghanistan, American or Soviet forces were directly involved in the fighting. In other cases, the two superpowers competed against each other by supporting a side in the conflict most aligned with them ideologically or most willing to accept their influence and assistance.

This war by proxy between the two superpowers also played out in the Middle East. The United States, which had been the first country to recognize Israel as an independent nation (followed shortly by the Soviet Union), was a strong supporter of the Jewish state during the Cold War. Some Arab countries enjoyed relatively cordial relations with both superpowers and some were aligned with the United States. The Soviet Union, which had a long tradition of Jewish persecution, forged a loose alliance with several of the Arab countries, including Egypt, Iraq, and Syria. Some of the countries in the Eastern

European Soviet bloc, especially Czechoslovakia (now the Czech Republic and Slovakia), also aided the Arabs. The Soviets also espoused sympathy for the Palestinians. Along with other Arab governments, the Soviets supplied the Palestinians with tools needed to stage terrorist attacks against Israel.

Religion was not at the heart of Arab-led terrorism during the Cold War; rather, it was Palestinian and Arab nationalism and unity, often referred to as Pan-Arabism. Israel was seen as an example of Western imperialism planted directly in the heart of the Arab world. The existence of a sovereign Jewish state also exacerbated the centuries-old rivalry between Jews and Arabs. Fighting against Israel was a fight against Zionism and a fight against Western imperialism. The Palestinian Arabs, thousands of whom were displaced as a result of the Jewish state, viewed the land of Israel as rightfully theirs. However, many world governments saw the existence of the nation of Israel as a haven against Jewish persecution, and many of the world's Jews and Christians saw the Jews as possessors of divine and historical claims to the land of Israel.

As much as many of the Arab states and the Palestinians detested Israel's existence, they eventually realized that the Jewish state was far superior to them in conventional warfare. With American and Western European backing, Israel quickly established itself as the top military power in the Middle East. A series of humiliating military defeats, starting in 1948 when Israel was attacked by armies of Syria, Iraq, Egypt, Lebanon and Jordan, led the Arabs to this conclusion. In almost every instance of direct conventional warfare, Israel single-handedly defeated several Arab armies at once. The Israelis defeated the armies of Egypt, Jordan and Syria in the Six Day War of 1967, and they defeated Egypt and Syria in the 1973 Yom Kippur War, even though Egypt was able to claim some success. No direct Arab military attack has occurred since.

In 1958, under Nasser's leadership, the nations of Egypt and Syria merged to form the United Arab Republic, but the partnership did not last; the United Arab Republic dissolved in 1961, but Egypt referred to itself as the United Arab Republic until 1971. Pan-Arab nationalism began to fail in the 1970s due in part to the 1970 death of its founder, Egyptian President Gamal Nasser. Arab unity was not achievable for many reasons, tribal and regional differences among them, and because Arabs have never been one unified group.

Among the Arab leaders to recognize the failure of Pan-Arabism and to officially acknowledge the existence of a Jewish state was Anwar Sadat, President Nasser's successor. Nasser severed Egypt's relations with the Soviet Union, embraced the United States as a partner, and in 1977 made the radical step of making peace with Israel, personally visiting Israel, and opening up formal diplomatic relations with the Jewish state. While Sadat's gestures toward Israel earned him worldwide acclaim and a Nobel Peace Prize, it infuriated many radical Muslims, including some in his own country, such as the Muslim Brotherhood, which included a physician named Ayman al-Zawahiri, who would be a principal founder of another terrorist organization called Egyptian Islamic Jihad, and would later serve as second-in-command to Osama bin Laden and Al Qaeda.

The Palestinians

Meanwhile, the nationless Palestinians, because they lacked the infrastructure to mount an open military campaign against Israel, and because their leadership included a substantial number of corrupt hacks and cold-blooded brutes, used terrorism to attract attention to their cause. With aid from their fellow Arabs and the Soviet Union, the word "Palestinian" became synonymous with "terrorism." Like most of their Arab brethren, most Palestinians are Muslims, but it was nationalism, not religion, that was the primary motive for acts of terrorism.

Even though many Arab leaders publicly voiced support for the Palestinian cause, this did not always translate into friendly relations between Arab governments and Palestinian leaders. Some of the bloodiest incidents in the Middle East during the Cold War were the result of rivalries between Palestinians and Arab leaders like King Hussein of Jordan and President Haffez Assad of Syria, both of whom were probably responsible for more Palestinian deaths than the Israelis, and both of whom were distrustful of Palestinian leaders who operated in their territory.

One of the major obstacles to peace in the Middle East is the question of the Palestinians, who claim rights to the land of Israel based on historical and religious grounds. Displaced and disenfranchised by the establishment of the Jewish state, and often treated as second-class citizens by their fellow Arabs, Palestinians were effectively without a voice for the first decade of Israel's existence. The strongest and most prominent voice of Palestinian nationalism came in the person of Yasser Arafat.

Yasser (or Yasir) Arafat was born in 1929, probably in Cairo, Egypt. A civil engineer by training, Arafat joined the Muslim Brotherhood and was president of the Union of Palestinian Students as a young man. He served in President Nasser's Egyptian Army in the 1950s. Later he journeyed to Kuwait where he established a business and founded Fatah, which would later become the military wing of the Palestine Liberation Organization (PLO), and which remains in force today. Although Arafat's name is associated the PLO, Arafat was not its founder. The PLO was founded in 1964 under the sponsorship of the Arab League. Its first chair was Ahmad Shuqeiri, a Palestinian diplomat. Arafat assumed control of the PLO in 1969.[16] For the next 25 years Arafat and the PLO would be seen as the face of Middle East terrorism. Hijackings and attacks against Israelis of all sorts were trademark PLO tactics.

The 1972 Olympics marked the arrival of Palestinian terrorism on the world stage. On September 5, 1972, eight Palestinian terrorists invaded the Olympic village in Munich, West Germany, and stormed the building housing the Israeli Olympic contingent. One Israeli athlete and one coach were killed in the raid. The terrorists demanded the release of 200 Palestinian prisoners in Israel. Nine other Israelis were taken to the airport as hostages. A botched rescue operation headed by the West Germans resulted in the merciless slaughter of all of the Israeli athletes, and the killing of a West German police officer and five of the terrorists.[17]

The Munich Massacre was planned to generate as much media attention as possible. The entire world watched the massacre unfold on television as media sources were already on hand in plentiful supply to cover the Olympics. In fact, the person who broke the news to Americans watching the event unfold on television was not a news journalist, but Jim McKay, an ABC sports journalist. The Munich

Massacre brought the problem of Arab terrorism and the Palestinian cause into the living rooms of people throughout the world, not just the Middle East.

The Revolution in Iran

Although it is difficult to rank events that created modern terrorism in any specific order of importance, the revolution in Iran in the late 1970s stands as one of, if not the most important, events in the history of twentieth-century terrorism. Iran was under the rule of Shah Reza Pahlavi, part of a monarchical dynasty that had ruled Iran for most of the twentieth century. Reza Pahlavi himself had ruled since his teen years. The Shah's regime grew increasingly out of touch with the Iranian populace. He was also unpopular because of human rights abuses, close ties to the West, and a feeling that his regime did not reflect true Islamic governing principles. In 1953, while the young Shah was in the twelfth year of his rule, the United States assisted in the overthrow of Mohammad Mosaddeq, an ardent Iranian nationalist who opposed Western intervention in Iran. Mossaddeq's overthrow was another blow to Iranian-Western relations. Coupled with the perception that Western values were increasingly encroaching on Iranian and Muslim culture in general and corrupting Muslim women and the youth, many Iranians nursed a strong hatred of the West and the United States in particular as the decades wore on.

One of the key figures of the twentieth century was Ruhollah Khomeini, an Iranian Shiite Muslim "Ayatollah" who was a dedicated foe of Iran's Shah and what Khomeini viewed as Western, and especially American, interference and corruption of Islamic culture. Khomeini and those who followed him were vocal advocates of government by his interpretation of pure Islamic law. Khomeini came to be synonymous with fundamentalist Islamic government. Unlike some other leading figures identified with Muslim militancy, Khomeini did not identify with communist-like beliefs and had almost as much disdain for the Soviet Union, Iran's long-time enemy, as he did for the United States, which he labeled "The Great Satan." Not being an Arab, and in fact looking askance at Arabs in general, Khomeini and his supporters did not identify with the gradually deteriorating Pan-Arabism that was so dominant in previous decades.

Khomeini and his fellow mullahs came to power in 1979 with the overthrow of an ailing, cancer-ridden Shah Palavi. When the former Shah went to the United States to seek medical treatment, a group of Iranian militants stormed the American embassy in Tehran on November 4, 1979, taking all of those in the embassy hostage. For 444 days, the United States and President Jimmy Carter were humiliated on the world stage, an embarrassment exacerbated by a failed rescue attempt on April 24, 1980. Most of the black and female hostages were released a few days after the takeover and one hostage was released a few months later for health reasons. The remaining 52 hostages remained in captivity until January 20, 1981, the day Ronald Reagan was inaugurated as President. Despite condemnation from governments around the world, Iran relished its role as a new power broker in the Middle East and the Muslim world, and it was emboldened to expand its influence and hegemony in the Middle East, Europe, Africa, and Southern Asia.

The Islamic government of Iran, headed by Ruhollah Khomeini, became emboldened by the American failure to deal with the hostage crisis of 1979–1981. When Iran was attacked by Saddam Hussein's Iraqi army in September 1980, many, including Saddam Hussein, believed that the new, unstable Islamic regime would not be able to hold the government together and that Iraqi influence and territory would expand as a result. However, the Iran-Iraq War, which lasted until 1988, did not produce a clear winner, but it did demonstrate to the world that the Iranian government could function while under attack from hostile forces, and the Islamic government not only continued to function, but became motivated by their successful resistance against their ancient Arab enemy.

During the 1970s and 1980s, Soviet influence experienced a decline in the Arab and Muslim worlds, in part due to its invasion of the Muslim nation of Afghanistan, which is discussed in the next section. While the Cold War was gradually winding down, and while Soviet influence and support for their Arab allies was slowly evaporating, the popularity of fundamentalist Shiite Islamic extremism, the kind practiced in Iran, was spreading. Some secular Arab leaders like Saddam Hussein of Iraq and the PLO's Yasser Arafat, who had in years past had seldom demonstrated any sense of devotion to Islam as part of their leadership philosophy, began wearing a religious façade in public, making it a point to be seen praying, attending mosques and speaking with noted religious leaders, even though they were adherents of the Sunni branch of Islam, unlike most Iranians, who were Shia Muslims. Despite their ethnic and religious differences, Yasir Arafat and the PLO managed to forge friendly ties with Iran. Even though Saddam Hussein hated his non-Arab Iranian neighbors, he was beginning to see that fundamentalist Islam, not Arab socialism, was key to attaining influence in the Middle East and the Arab and Muslim worlds.

The Soviet Invasion of Afghanistan

On December 24, 1979, the Soviet Union shocked the world by invading Afghanistan and installing a pro-Soviet government. Since its inception, and especially after World War II, the Soviet Union had aggressively sought to expand its influence throughout the world, especially in or near its bordering countries, but most of the world's governments assumed that the days of brazen Soviet expansionism by military means was a thing of the past. The Soviet Union in and of itself was an artificial confederation of states dominated by Russia. The Soviet Republics shared few cultural commonalities, so many Soviet citizens, especially those in predominantly Muslim republics such as Kazakhstan, Uzbekistan, and Turkmenistan never felt a strong sense of allegiance to the Soviet Union. The Warsaw Pact, which bound the Soviet Union into a forced alliance with its satellite countries in Eastern Europe, was unpopular in most of those countries too, especially Czechoslovakia, Hungary, and Poland. Much of this Soviet aggression was the work of a series of World War II-era Communist leaders, who, by the early 1980s, were beginning to age and die, which was symptomatic of the Soviet empire itself.

The invasion of Afghanistan signaled the unraveling of the Soviet empire. Muslims from around the world, especially the Middle East and South Asia, including some of the member Soviet Republics, came to Afghanistan to assist in the jihad against the infidel Soviets. Afghan rebels and their non-Afghan

Muslim allies received considerable assistance from other governments, including some of the Arab countries, but also from Pakistan and the United States. After years of fighting, with the Soviet empire and its economy crumbling, its population weary of war, and its Cold War leadership dying and being replaced by younger and more pragmatic leaders like Mikhail Gorbachev, the Soviets withdrew from Afghanistan in February of 1989.

The successful though bloody victory over a superpower at the hands of a rebel force was viewed by many Muslims as a miraculous victory for Islam over the infidels. Afghan rebel forces and their non-Afghan Muslim allies were emboldened by the withdrawal of Soviet forces from a Muslim country. Unfortunately, much of the foreign assistance that was so helpful to the defeat of Soviet forces dried up after the Soviet withdrawal, creating a huge power vacuum in Afghanistan and leaving the country with virtually no centralized government. This vacuum opened the door for a bloody takeover of the Afghan government by a religious group called the Taliban, a group of ignorant, uneducated Muslim fanatics who knew little about governing, but who were determined to impose a strict, reactionary form of Islamic law on the population of Afghanistan. Many of those supported by the United States against the Soviets now turned their animosity toward their former benefactors, including the United States, other Western governments and secular Muslim governments that enjoyed good relations with the West, including Saudi Arabia, Jordan, and Egypt.

The Sadat Assassination

While Muslims were fighting the Soviets in Afghanistan, Egyptian President Anwar Sadat paid a high price for making peace with Israel and increasing Egypt's ties with the United States. On October 6, 1981, Sadat was assassinated by soldiers in the Egyptian army (who were allied with the Muslim Brotherhood and Egyptian Islamic Jihad) while he sat in a reviewing stand. More than 300 defendants were tried for the Sadat assassination, including Ayman al-Zawahiri, the eventual second-in-command in Al Qaeda, and Omar Abdul Rahman, a blind Egyptian cleric who would later be involved in the 1993 World Trade Center bombing.[19]

Although the Sadat assassination was viewed as a tragedy in the West and in the United States, many in the Arab and Muslim world celebrated his death. Sadat was viewed by most of his fellow Arab leaders as a traitor; in fact, only one Arab government sent an official representative to his funeral. Sadat's assassination also motivated Muslim extremists, who had just begun fighting the Soviets in Afghanistan and were about to increase their presence in Lebanon.

Israel Invades Lebanon

Since its creation, the nation of Israel has been surrounded by hostile neighbors. Even though the governments of Israel have enjoyed a relatively peaceful relationship with its southern and eastern neighbors, Egypt and Jordan, Israel still faces threats from many in those countries. Israel has long had

The Continuing Role of Interpol

Among the international organizations involved in global law enforcement, few are as active as the International Criminal Police Organizations, known since 1956 as Interpol. Established in 1923 as a private organization of senior police officers in Europe, the International Criminal Police Commission served as an informal place for the exchange of information concerning criminals who conducted their activities across national borders. With the fall of the Austrian Republic into Nazi control in 1938, the commission ceased to function at the international level and many of its records were apparently destroyed. At the end of World War II, the commission's headquarters was established at Paris, and with the aid of a French government loan, Interpol moved into a new building at Saint-Cloud in 1967.

Shortly after Interpol was situated in its new headquarters, the outbreak of the Cold War resulted in the withdrawal of Eastern Bloc nations from membership. The United States, represented by the Federal Bureau of Investigation, withdrew from membership after 1950. This was in protest of Interpol's identification of individuals who hijacked a Czechoslovakian airliner for a flight to freedom in 1949. The United States resumed membership in 1958, with the Treasury Department and its Secret Service being the main participants in Interpol activities.

Throughout its history, Interpol has been reluctant to serve as a channel for distributing information concerning "political" crimes. Most controversial has been its refusal to assist in the location of Nazi war criminals, many of whom had been convicted in absentia at Nuremberg. On the other hand, Interpol's refusal to assist communist Cuba in searching for President Fulgencio Batista's regime leaders has been widely acclaimed by western nations and the United States. In 1988, Interpol reviewed its policy concerning Nazi war criminals and has since been active in assisting nations seeking to apprehend them. Interpol had also refrained from participating in the investigation of terrorism, deeming terrorist activities "political crimes." However, the 1984 Luxembourg General Assembly of Interpol members agreed to assist in investigating terrorist acts, but provided that the decision to participate would be made on a case-by-case basis.

Unfortunately, the degree of national cooperation with Interpol, as well as compliance with Interpol requests for information, is shaped by the member nations' divergent concepts of criminal law enforcement and by their assessment of the appropriate scope of police activity. In the past two decades, growing international law enforcement needs have increased the value of police cooperation across departmental and national boundaries. That is particularly the case in regard to drug trafficking, as many nations (including the United States) have stationed police officials in other nations, hoping to halt the illegal trade in drugs. Interpol has exercised a growing role in the exchange of police intelligence, in the capture of most-wanted criminals, and in the location of missing persons. Cross-national efforts to control drug trafficking, solve terrorism cases, and enforce laws against smuggling have also increased the frequency of using the extensive files maintained by Interpol. The organization has thus expanded from a European institution to a police intelligence activity that has global recognition and impact.

Policing international criminal activities has a long way to go before it will proceed without disagreement and tension. Professor Malcolm Anderson points out that without a harmonization of criminal law procedures—such as the U.S.-British common rules concerning drug trafficking and policing this threat—there is a limited amount of success that can be achieved through international organizations such as Interpol. There is also strong American and British skepticism concerning the centralized police systems of Europe, which place inordinate emphasis on the assembly of large dossiers of information on all citizens, not only those who have criminal records.

The reluctance of Interpol to deal with terrorism prior to 1984 has resulted in the rise of other institutions for the exchange of international police information. Under the category of European Political Cooperation activity, sponsored by the European Union, three so-called Trevi groups have emerged as focal points for inter-European cooperation in law enforcement. Trevi 1 was organized in 1975 to establish a secure communications system that would relay information concerning terrorists and their activities. Trevi 2, established later, is dedicated to the exchange of information on police methods and systems, and Trevi 3 deals with approaches to serious crime other than terrorism.

an openly hostile relationship with its northeastern neighbor, Syria, and for decades Syria has been a dominant force in the government of Lebanon, Israel's neighbor to the north. From its inception, Israel has had to deal with threats from Lebanon, whether outright military confrontation, infiltration of terrorists from Lebanon into Israel, or having to contend with bombs and rockets launched into Israeli territory from Lebanon. In 1982, Israel invaded Lebanon in reaction to attacks from Lebanese territory. The objective was to neutralize hostile forces in Lebanon and to create a buffer zone sufficient to shield northern Israel from encroachments and rockets launched from Lebanon.

The Israelis succeeded in creating the buffer zone, occupying southern Lebanon for more than 20 years. However, the invasion of Lebanon brought an unintended consequence. Palestinian attacks from Lebanon had always enjoyed the approval of many Arab and Muslim countries, and this support only increased with the Israeli incursion into Lebanon. Iran, buttressed by its humiliation of the United States because of its takeover of the American embassy, increased its support to the Palestinians in southern Lebanon. With aid from Iran and Syria, Hezbollah (the Party of God) came into being and still lives on as an important player in Middle East politics and as a nemesis of Israel. Although some countries, including the United States, Canada, and Israel, classify Hezbollah as a terrorist organization, other countries, especially Arab and Muslim countries, view Hezbollah as a legitimate resistance organization.

Hezbollah is not only a menace to Israel and its political opponents in Lebanon, it was responsible for one of the deadliest attacks against Americans in the 1980s. President Ronald Reagan had ordered a contingent of Marines to be stationed in Beirut, Lebanon, as part of an international peacekeeping force. On October 23, 1983, a car bomb driven by a suicide bomber rammed into the Marine barracks in Beirut, killing 241 American service personnel. The Marine bombing was one of many incidents that occurred in Beirut during the 1980s that defined the city as one of the most unsafe places on earth. At that time Hezbollah established a presence, which remains to this day, in Lebanon's Bekaa Valley.

International Cooperation in the Suppression of Terrorism

On December 21, 1988, Pan Am Flight 103 exploded in the sky above Lockerbie, Scotland. All the passengers and crew, totaling 276 persons, were killed, and the village of Lockerbie lost residents as airplane fragments crashed to the ground. At the time it was known that Pan Am Flight 103 originated in Malta, and had made intermediate stops at Frankfurt and London before it took off for New York City. Forensic evidence indicated that the bomb particles were of Libyan origin, and further investigation led to the identification of Ali Mohmed al-Megrahi and Al Amin Khalifa Fhimah and implicated them in the placement of the bomb. After the imposition of economic sanctions against Libya, the government of Colonel Quaddafi agreed to deliver the two men to international authorities, provided that they would be tried in the Netherlands before Scottish judges. The trial lasted nine months and was held at Camp Zeist, a former NATO base in the Netherlands. Understandably, after 12 years the evidence against the defendants was limited, and their identification as persons involved in the preparation of a bomb and its placement in Flight 103 was uncertain. Nevertheless, the court found al-Megrahi guilty and sentenced him to life imprisonment, which was mandatory under Scottish law. After imposing the sentence, the judges told Megrahi that they would recommend he be eligible for parole in 20 years. This was the strongest penalty that was available to the court. The second defendant was acquitted, the only evidence against him being diary entries indicating a meeting with Megrahi and his acquisition of suitcase luggage tags. Unlike the record on Megrahi, there was no connection established between Fhimah and the Libyan intelligence service. Two prosecution witnesses, a Libyan secret service informant and a Swiss electronics firm owner, were considered to have given unreliable and untruthful testimony that the judges discounted. On the other hand, Megrahi had dealt with the Swiss electronics firm Mebo, and it was a Mebo timer that triggered the Lockerbie bomb. Within the suitcase that held the bomb, investigators found clothing very likely purchased by Megrahi at a clothing store in Malta.

The conclusion of the Camp Zeist trial provides insight into the utility of criminal justice systems as instruments to combat terrorism. Newspaper comments of the day suggested that formal rules of evidence and procedure make it difficult to secure justice against accused terrorists. Others argued that the measured use of military force is a more appropriate and more effective response to discourage nations that adopt terrorism as a national policy. Undoubtedly the sparsity of the evidence and documentation, coupled with Libya's initial refusal to surrender those accused of this crime, were partially responsible for the acquittal. On the other hand, one of the accused was tried, found guilty, and sentenced to life imprisonment with a possibility of parole after 20 years.

Oklahoma City

On April 19, 1995, a truck bomb exploded at the Murrah Federal Building in Oklahoma City. The final death toll was 168 and thousands were injured. Some of the dead were preschool children who had been dropped by their parents at the building's day care center. Suspicions immediately abounded that the bombing had all the earmarks of Middle Eastern terrorism. Many Americans were surprised when, a few days later, a decorated U.S. Army Persian Gulf War veteran was linked to the bombing. Timothy McVeigh was stopped by an Oklahoma State Trooper for driving without a license plate; he was also found to have an illegal weapon. He was in jail for these offenses when evidence was discovered that linked him to the bombing.

McVeigh and a co-defendant, Terry Nichols, were charged with the bombing. Both had loose affiliations with right-wing anti-government militia groups in the Midwest, but the leaders would claim no formal affiliation with McVeigh or Nichols. McVeigh was angered about an event that occurred in Texas one year earlier. In February of 1994, Alcohol, Tobacco, and Firearms agents raided the Waco compound of a religious cult called the Branch Davidians, who were maintaining a cache of illegal weapons; they were suspected of other crimes as well. On April 19, when federal agents attempted another raid, the compound was burned to the ground, killing all those inside. The Waco incident, and the federal government's involvement in it, proved to be a lightning rod for right-wing opponents of the federal government. Exactly one year after the Waco raid, the Oklahoma City bombing was carried out in revenge.

McVeigh was also influenced by an underground novel called *The Turner Diaries*. Written under the pseudonym Andrew Macdonald, *The Turner Diaries* was the work of William Luther Pierce III, who had formal ties to several white supremacists and right-wing groups throughout the Midwest, including the American Nazi Party. Pierce, who had earned a Ph.D. from the University of Colorado, was a former physics professor at Oregon State University. In his novel, which had been circulating in non-legal circles for almost 20 years, Pierce's protagonist, Turner, was involved in the blowing up of FBI headquarters, the killing of non-whites and Jews, and the persecution of white women who had had sex with non-whites.

McVeigh and Nichols were convicted of the bombing. McVeigh was sentenced to death and executed on June 11, 2001. Nichols was convicted and sentenced to life imprisonment. Michael Fortier, a former Army buddy of McVeigh's who had prior knowledge of the plot, testified against the pair and received a 20-year prison sentence.

The Oklahoma City bombing reminded Americans that terrorism was not just a radical Muslim phenomenon. Right-wing terrorism was neither created nor did it die with Oklahoma City, but no right-wing terrorism of this scale had ever been perpetrated. In fact, until 2001, the Oklahoma City bombing was the deadliest terrorist attack in U.S. history on American soil.

Sources: J.R. White, *Terrorism and Homeland Security*, Fifth Edition (Belmont, CA: Thomson Wadsworth, 2006; A. Macdonald, *The Turner Diaries*, Second Edition (Fort Lee, NJ: Barricade Books, 1996, original publication date 1994; Wilson Web Biographies, http://vnweb.hwwilsonweb.com.jproxy.lib.ecu.edu, accessed March 5, 2008.

Al Qaeda and Osama bin Laden

One of the Arab foreigners who came to Afghanistan to fight the Soviets was Osama bin Laden. Born in 1957 in Saudi Arabia to a family of Yemeni immigrants, bin Laden was raised amid a life of privilege.[20] His father, the father of 57 children (Osama bin Laden was his seventeenth), was a millionaire construction magnate in Saudi Arabia. Osama bin Laden graduated from King Abdul Aziz University in 1979 with a degree in civil engineering and, like many other young Muslims from around the world, went to Afghanistan to join the fight against the Soviets.[21]

At least one account of the Soviet-Afghan conflicts states that bin Laden's role in the conflict has been greatly exaggerated by bin Laden and his minions. George Crile states, "It is ironic that a man who had almost nothing to do with the victory over the Red Army, Osama bin Laden would come to personify the power of the jihad."[22] Whether deserved or not, bin Laden emerged from the war a hero in the eyes of many Muslims, and he used this reputation to build his organization and his power base far beyond Afghanistan. After the Soviets withdrew from Afghanistan, bin Laden worked toward solidifying his control over the remnant Muslim factions that had worked together in Afghanistan. He had his chief partner, an Egyptian named Abdallah Azzam, assassinated on November 24, 1989, leaving him in complete control of his emerging network of Muslim fighters, which by this time had acquired a name, Al Qaeda, or "the base."[23]

On August 2, 1990, Iraq invaded Kuwait, its tiny oil-rich neighbor to the south. Iraqi President Saddam Hussein announced that Iraq was annexing Kuwait, claiming that Iraqis had a legitimate claim to the tiny emirate, and that Kuwait was an illegitimate nation created to serve Western oil interests. Kuwait and the Western countries, including the United States, had enjoyed a mutually beneficial relationship for many years. Almost immediately, U.S. President George H.W. Bush announced, along with the leaders of many other countries, that the Iraqi occupation of Kuwait would not be tolerated. World leaders became even more nervous when it was feared that Iraqi forces would invade Saudi Arabia. Soon after it became apparent that Iraq would not withdraw from Kuwait, the United States and other countries, including Great Britain, began to amass forces to prepare to extract Iraqi forces from Kuwait, while President Bush recruited other countries into its coalition. Most notable was the fact that Egypt and Syria, two Arab countries, joined the coalition against Saddam Hussein.

Now back in his native Saudi Arabia, and regarded by many there as a war hero, bin Laden and his followers saw that Saudi Arabia, or as he called it, the "Land of the Two Holy Sanctuaries" (Mecca and Medina), was under siege by Iraq. He let it be known to the Saudi monarchy that he and his forces were ready and able to go to war against Saddam Hussein's forces.[24] Bin Laden viewed Saddam Hussein's government as non-Islamic and as too secular, a by-product of Gamal Nasser's Arab Socialism. In fact he often referred to Iraq and Syria as "communist" countries. Bin Laden also saw the Iraqi invasion as an opportunity to expand his own influence in the Muslim world. In bin Laden's view, Muslim rebels, under divine inspiration and guidance, had succeeded in ridding Afghanistan of the Soviets; therefore, they would be able to extract the Iraqis from Kuwait and protect the "Land of the Two Holy Sanctuaries" from Hussein's aggression. According to bin Laden, ridding the Arab world of Saddam Hussein's menace was a task best left to fellow Arabs.

Bin Laden was incensed when the Saudi monarchy turned to Western governments to help them with their battle against Iraq. He was outraged even further when the Saudi government allowed Western military forces to establish operations in Saudi Arabia from which the invasions of Kuwait and Iraq would be launched. In bin Laden's view, Saudi Arabia was sacred ground reserved for Muslims only. The Saudi monarchy, which was charged with guarding and protecting Islam's holiest cities, committed the ultimate act of apostasy by allowing infidel Western forces on its soil. This was compounded when Saudi Arabia allowed foreign forces to remain there after the brief war's end.

The 1993 World Trade Center Bombing

The devastation of September 11, 2001 was so enormous that it has eclipsed the first attack on New York City's World Trade Center. On February 26, 1993, terrorists detonated a car bomb beneath the North Tower of the World Trade Center. The objective was to create enough of an impact to cause both towers to fall, which would have resulted in thousands of casualties. The terrorists did not succeed in bringing down the towers, but on that occasion six people were killed and more than 1,000 were injured.

Several men were eventually convicted of the World Trade Center bombing, including Ramzi Yousef, the nephew of Khaled Sheikh Mohammad, who would also be the chief architect of the September 11, 2001, attacks. Also convicted was Omar Abdul Rahman, who had been tried and eventually released in the assassination of Egyptian President Anwar Sadat several years earlier. Despite the successful prosecution and prison sentences imposed on many of the perpetrators of the World Trade Center bombing, the first attack was merely a trial run for the much more devastating attacks that followed eight years later.[25]

The 1996 Khobar Towers Bombing

By 1994, bin Laden's activities brought him under the watchful eye of the Saudi government, and even many members of his family had disassociated themselves from him. Bin Laden was expelled from Saudi Arabia in 1994 for subversion. He went to Sudan, where he organized terrorist training camps. Bin Laden was expelled from Sudan in 1996 and went to Afghanistan, where his wealth and popularity enabled him to stay as the "guest" of the ruling Taliban.[26] Al Qaeda was not the only entity that opposed the American presence in Saudi Arabia. An eight-story building in Khobar, Saudi Arabia, which housed foreign (including American) service personnel and their families, was the target of a truck bomb on June 25, 1996. Nineteen American service personnel were killed, as was one Saudi. Hundreds of soldiers of different nationalities were injured.

More than a dozen men were indicted for the bombings, but the exact identity of those behind the Khobar bombing is still unclear. Suspicion has centered on Hezbollah, who receive assistance from Syria and Iran, on high-level Iranian government officials, and on Al Qaeda.[27] Given the animosity between the Shia Muslim government of Iran and the Sunni Muslim-dominated Al Qaeda, it is unlikely that there was a partnership between the two.

Al Qaeda Accelerates: The 1998 Embassy Bombings and the USS Cole

In February 1998, bin Laden and Ayman al Zawahiri, by now bin Laden's second-in-command, arranged for a fax to be sent from Afghanistan to an Arabic newspaper in London. In the fax, they claimed that the United States had declared war against God and his messenger (the prophet Mohammed). Bin Laden stated that it was the duty of all Muslims to kill any American anywhere on earth. Bin Laden declared war on the United States, and that any and all Americans were fair targets because by paying taxes they supported the actions of the U.S. government. In May 1998, in an interview from Afghanistan with ABC Television's John Miller, bin Laden stated, "We do not have to differentiate between military or civilian. As far as we are concerned, they are all targets."

Bin Laden proved true to his word less than three months after his interview was broadcast. On August 7, 1998, car bombs exploded at the United States Embassies in Nairobi, Kenya and Dar Es Salaam, Tanzania. The attacks were practically simultaneous. The Nairobi bombing claimed 212 lives, most of them Kenyans, and the Tanzania bombing claimed 11 lives.

More than a dozen members of Al Qaeda were indicted for the bombings, including Osama bin Laden and Ayman Al Zawahiri. Some have been killed; some were sentence to life in prison; some were held in prisons outside the United States; and others, including bin Laden and Zawahiri, were still at large as of this writing.

Al Qaeda's attacks against the United States were growing bolder and more deadly. In response to the East African bombings, President Bill Clinton authorized Operation Infinite Reach, in which cruise missile attacks were launched in Sudan and Afghanistan. The Sudan raid demolished a pharmaceutical plant, but little real damage was done. In fact, many have argued that the token response by the United States military only demonstrated that America was a paper tiger, lacking the will or capability of crippling Al Qaeda. Bin Laden's fame grew; he even earned praise from the Iraqi government that he despised, and from the ruling Taliban in Afghanistan, who had welcomed bin Laden as their "guest." Taliban hospitality was motivated in large measure by Al Qaeda money. Despite the sometimes prickly relations between the Taliban and Al Qaeda, the two entities enjoyed a symbiotic relationship. Al Qaeda needed a safe haven for its operations and the Taliban needed Al Qaeda's money, and neither entity could rid itself of the other.

Al Qaeda was not finished. On October 12, 2000, the *USS Cole* was docked in the port of Aden, Yemen. Yemen, long a hotbed of anti-Americanism, radical Islam, and virtually lawless tribal bands. On October 12, suicide bombers aboard a small boat rammed the *Cole*. The attack killed 17 American sailors and wounded 39 others. The alleged mastermind of the *Cole* bombing was killed on November 4, 2002 by a predator drone operated by the CIA.

September 11, 2001

On the morning of September 11, 2001, there was nothing to suggest that the day would be eventful. Before the morning was over, this would change drastically. At 8:46 A.M., a plane crashed into the North Tower of New York's World Trade Center. At first, many assumed that the crash was simply the result

Why Do They Hate Us?

"Why do they hate us?" was one of the most frequently repeated questions in the aftermath of September 11, 2001. There is no simple answer. Thomas Friedman of the *New York Times* states that many pockets of the Muslim world, especially the Arab world, feel humiliated by the Western world, left behind in technological advances, democratic advances, and quality of life in general.[32] Many Arabs and Muslims see the existence of the nation of Israel as a symbol of that humiliation. In addition to the historic enmity that has existed between Jews and Muslims, many modern Muslims blame Western Europe and the United States for the existence of Israel. Some believe that it is Western guilt over the Holocaust and over the persecution that Jews suffered in Europe for centuries that resulted in the establishment of Israel. In other words, today's Palestinians are paying the price for Western guilt over treatment of the world's Jews. In addition, the perception is that most Israelis enjoy a democratic government and live lives of comfort, at least in relation to the vast majority of Arabs and Muslims who live in abject poverty under the oppressive thumb of dictatorial governments supported by Western oil money.

Another reason offered for the enmity between the Muslim world and the West is religious. Some fundamentalist Muslims believe that modern culture is simply not compatible with Islamic teachings, thus the Taliban's insistence on the banning of televisions, radios, mirrors, and anything that is not specifically endorsed by the Quran. Shmuel Bar refers to this as lacking the tools of anachronization, the inability to distinguish what was right in a bygone time versus what is right or wrong in modern times, with the only solution being to try to make life as much like it was in the time of Mohammad as possible.[33]

There are a few in the Muslim world, especially among the Shia, who are awaiting or trying to facilitate the appearance of the "hidden Imam," which is akin to Christian apocalyptics who are awaiting the return of Jesus Christ to earth. This hidden Imam will bring Allah's kingdom to earth and vanquish the world of Allah's enemies. This belief, coupled with Quranic commands to wage war against the enemies of Allah who refuse to convert to Islam, provide the theological justification for hatred of non-Muslims, although many Muslims view such an interpretation as a perversion of the Quran. Whatever the motive, non-Muslims and moderate Muslims would be wise to remember that the most fanatical Islamists see themselves as operating on "God's timetable." The global jihad, or holy war, will last as long as the world is not completely surrendered to the will of Allah, a period that will last far beyond the life of anyone now living on Earth. If the jihad must last 1,000 years, so be it.

of pilot error, and that the plane was probably a small private jet. However, questions arose about the likelihood of such an error, given that the weather on September 11, 2001, in New York City was clear and sunny, a typical late summer day.

Unbeknownst to all but a few, the plane that crashed into the North Tower was American Airlines Flight 11, with 92 passengers and crew aboard. The flight had taken off from Boston and was bound for Los Angeles. It had broken off contact with the FAA a few minutes after takeoff. Even though the FAA knew that American Airlines 11 was inexplicably headed toward New York City, and even though they suspected that a hijacking was in progress, they were unsure if the plane that had crashed into the North Tower was indeed Flight 11. Within minutes the worldwide media had arrived to cover the crash, still unsure as to its nature.

Any doubt as to the cause of the crash was put to rest when at 9:03 A.M., a second plane crashed into the South Tower of the World Trade Center, in full view of the television media from around the world that have arrived to film the first crash. The second plane was United Airlines 175 with 65 crew and passengers aboard. It too was a Boeing 767 jet traveling from Boston to Los Angeles.[28]

The situation worsened as the day wore on. At 9:37 A.M., American Airlines Flight 77 (a Boeing 757), traveling from Dulles Airport in Washington, D.C. to Los Angeles, and carrying 62 passengers and crew, crashed into the Pentagon near Washington, D.C. Like the two Boston flights, it had been selected by the hijackers in part because cross-country flights carry more fuel than short flights, thus maximizing the damage inflicted by a crash.

At 10:02 A.M. a Boeing 757, United Airlines Flight 93, originating in Newark, New Jersey, and bound for San Francisco and then Tokyo, crashed into a field near Shanksville, Pennsylvania. The plane carried 44 passengers and crew. Hearing of the World Trade Center crashes through phone calls to friends and relatives on the ground, United 93 passengers had attacked the hijackers and stormed the cockpit, causing the plane to crash. It was determined later that United 93 was probably headed for either the White House or the United States Capitol. According to the makers of the film *United 93,* the pilot hijacker placed a picture of the United States Capitol on the control panel. United 93 failed to reach its intended target because it had taken off 25 minutes late, which threw off the hijackers' timing, allowing time for passengers on board to get word from the ground about what had happened in New York City and at the Pentagon.[29]

Ben Sliney was assuming his first day as Director of FAA Operations on September 11. After the World Trade Center attacks he ordered all commercial air traffic halted, and he ordered the grounding of all planes already in the air. The military was alerted and presidential authorization was given to shoot down any plane that would not comply with the order to land or that was thought to be hijacked.[30] Fortunately, this order did not have to be carried out.

To many Americans, life in the United States seemed surreal. Major League Baseball ordered all games postponed, in large measure because teams could not travel, but also because there was a fear of terrorist activity in large gatherings. News outlets covered the incident non-stop for days. Commercial air traffic was completely suspended for three days. Not since World War II had Americans experienced such an emotional and psychological jolt, and no event as traumatic and deadly had occurred on the soil of the American mainland since the Civil War. Even those who compared 9/11 with the

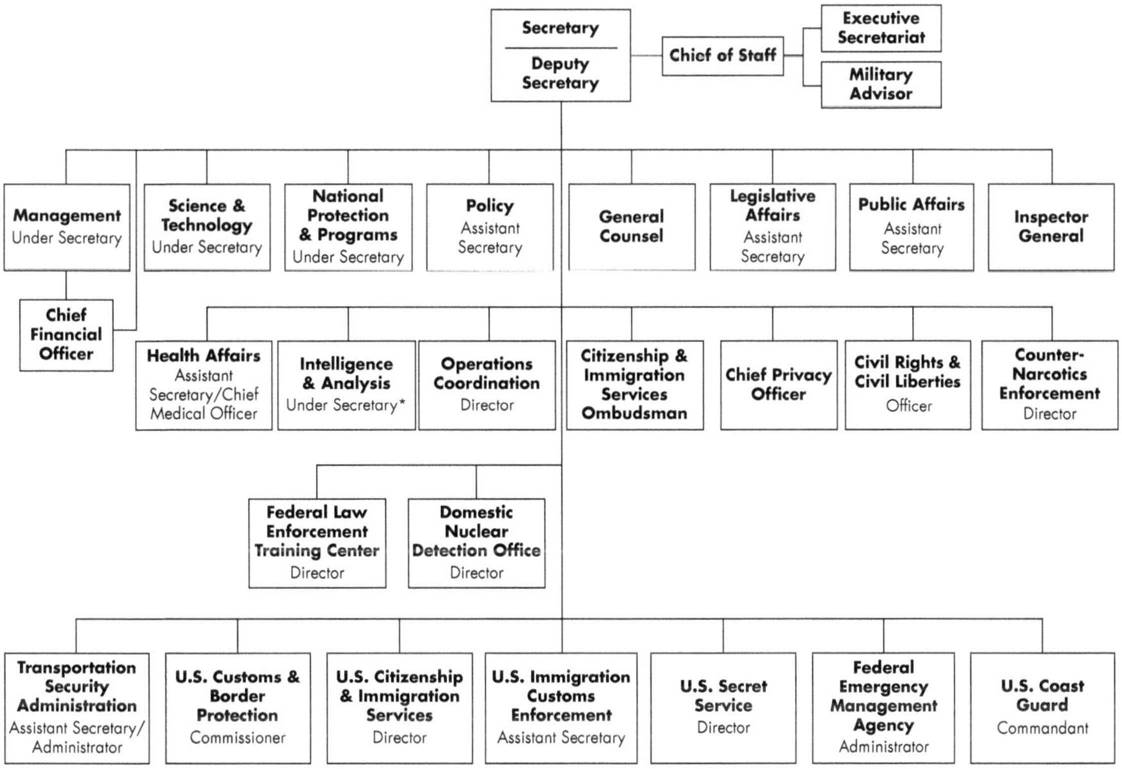

*Under secretary for Intelligence & Analysis. Title created by Public Law 110-53, August 3, 2007.

Pearl Harbor attack noted that as horrible as the Pearl Harbor attacks were, at least the Japanese had targeted a military installation, rather than thousands of unsuspecting civilians of all ages.

At the time of the World Trade Center attacks, President George W. Bush was in Sarasota, Florida, speaking to a group of schoolchildren. When he was notified of the first crash, Bush was told that it may have been an accident. When he was notified of the second crash by Chief of Staff Andrew Card, Bush wore the same look of shock and chagrin as many Americans; he knew that America was at war. After a day of being flown to several different airbases because it was believed that the White House might come under attack, Bush returned to the White House that evening, where he delivered a televised address, in which, referring to the Taliban, he stated that the United States would make no distinction between terrorist entities and the governments that supported them.

Nineteen hijackers carried out the 9/11 crashes. Fifteen were Saudi Arabian nationals and all were Middle Eastern or North African Muslims. Attention turned immediately to Osama bin Laden, Ayman al-Zawahiri and Al Qaeda. The attacks of September 11 claimed 2,974 lives, plus the lives of the 19 terrorists.[31] It thrust the United States into a shadow war against an enemy without official ties to a sovereign government, and it established the course of American foreign and defense policy for years to come. The attacks also had a profound effect on domestic issues in the United States, many of which are criminal justice-related.

The attacks of September 11 had been in the planning phases for at least five years. The hijackers spent several years simply traveling by airplane from place to place, acquainting themselves with the layout of various planes, the habits and nuances of flight crews, and the security systems at various airports. The hijackers each had their own role to play and had received extensive training in the role they were scheduled to play in the attacks. Some acted as muscle men, whose primary job was to take over the plane and immediately kill or subdue enough passengers and crew to cow the other passengers into submission. The primary weapons were box cutters, which the hijackers carried aboard the plane legally. One hijacker would attach a fake bomb to his chest, thus deterring the passengers from fighting back. The hijackers led the passengers to believe that the plane was merely headed to an airport. Thinking that the hijackers would release them pending some sort of ransom demand, the passengers aboard the New York City and Pentagon flights did not know that they were part of a suicide mission. Every plane had at least one hijacker designated to pilot it once the crew was removed or killed. These hijackers had received training at flight schools in the United States, amid a degree of curiosity from some in the flight schools when they learned that some of these individuals wanted to learn how to pilot a plane but not how to land it.

Although many terrorist experts had predicted that some sort of terrorist attack against American interests was in the planning stages, including the possibility of using commercial airliners as weapons, and Osama bin Laden had unabashedly stated in 1998 that he had declared war on the United States, the attacks of September 11, 2001, still caught America by surprise. Terrorism was not new to Americans, but mass murder on such a brazen and massive scale perpetrated on American soil by foreign interests shocked Americans to the greatest extent since the bombing of Pearl Harbor.

Shortly after the September 11 attacks, President George W. Bush's administration requested that the ruling Taliban in Afghanistan either expel Osama bin Laden and his top lieutenants or turn them over to American authorities. There was little optimism about this request and, as expected, the Taliban failed to comply. In October 2001, the United States and other allied nations invaded Afghanistan and deposed the Taliban, but its main leader, Mullah Mohammad Omar, eluded the allies' grasp, as did Osama bin Laden and Ayman al-Zawahiri. All three were believed to have taken refuge, though probably not together, in Pakistan or the borderless tribal regions between Afghanistan and Pakistan. Prior to the 9/11 attacks only three countries had formal diplomatic ties with Afghanistan, which was effectively a nation without a functioning government. Those three countries—Saudi Arabia, Pakistan, and the United Arab Emirates—severed relations with Afghanistan after 9/11.

The Present and the Future: Homeland Security

What implications does terrorism have for criminal justice today and in the future? Prior to September 11, 2001, most terrorist attacks in the United States were investigated and prosecuted much like any other crime. If American interests in other countries were attacked, they were sometimes dealt with militarily, sometimes through the civilian criminal justice system, and sometimes with a combination of military and civilian means.

In his nationally televised address to a joint session of Congress on September 20, 2001, President George W. Bush introduced a term that was new to many Americans, but has become a household term since that night. The term was "homeland security." So many Americans were swept up in both the fear and patriotic euphoria generated by 9/11 that they probably could not grasp the future significance of this portion of President Bush's speech:

> Today, dozens of federal departments and agencies, as well as state and local governments, have responsibilities affecting homeland security. These efforts must be coordinated at the highest level. So tonight, I announce the creation of a Cabinet-level position reporting directly to me, the Office of Homeland Security. And tonight, I also announce a distinguished American to lead this effort, to strengthen American security: a military veteran, an effective governor, a true patriot, a trusted friend, Pennsylvania's Tom Ridge. He will lead, oversee and coordinate a comprehensive national strategy to safeguard our country against terrorism and respond to any attacks that may come.[34]

Tom Ridge became the first Director of the Office of Homeland Security in October 2001. Following Congressional approval, the Office of Homeland Security was granted cabinet-level status, and on January 24, 2003, Ridge became the first Secretary of the Department of Homeland Security (DHS). Tom Ridge stepped down as DHS Secretary in February 2005 and was succeeded by Michael Chertoff, a former federal appellate court judge and prosecutor.

The attacks of September 11, 2001, altered the structure of the federal government; it affected state and local governments as well. Numerous state and local agencies began competing for DHS funds and lined up behind DHS directives to secure their localities against the threat of terrorism. In law enforcement circles, police agencies began implementing training for their officers, both in counter terrorism, recognition of terrorism indicators, and first responder training. Perhaps most significantly for the criminal justice system, the issue of information sharing was addressed. It turned out that numerous agencies had assembled information on some of the 9/11 hijackers and some of them were on terrorist watch lists both in the United States and other countries, but no mechanism existed so that these agencies could pool their information, and allow some centralized information specialists to "connect the dots" and foresee the plot. At least three of the hijackers had been stopped for traffic offenses not long before the 9/11 attacks. One was stopped twice in Florida in July; another was stopped in Virginia in August for speeding; a third was stopped in Maryland on September 9 for speeding.[35] Had the Maryland police officer had access to the information that the federal government had on this individual, namely that he was on terrorist watch lists in several countries, he may have been able to at least detain one of the hijackers, which may in turn have thwarted the entire operation.

DHS was assigned the monumental task of assembling information from the vast number of federal, state, and local agencies that may have information related to terrorism. Fusion centers, which operate throughout the country, are being used as one such means for assembling information. Information specialists have become front-line soldiers in the United States anti-terrorism fight just as much as soldiers, sailors, and law enforcement officers.

The impact of the homeland security focus has also been felt in higher education. Terrorism and its prevention have been areas of academic inquiry for many years, but the creation of the DHS and other homeland security initiatives have greatly increased this focus. In fact, an entirely new academic discipline is evolving. The study of homeland security, sometimes referred to by other terms such as security studies, is evolving in a manner similar to the manner in which criminal justice evolved in the 1960s and 1970s. Like criminal justice, homeland security is a combination of several disciplines, typically encompassing criminal justice, political science, environmental health, geography, and information management. Like criminal justice, which benefited greatly from the creation of a new government agency, the Law Enforcement Assistance Administration, homeland security study has benefited from the creation of the DHS. In years to come, some professors in criminal justice departments may find themselves as professors in departments of security studies, much as many sociologists, psychologists, political scientists and social work professors have found themselves in criminal justice departments rather than their native disciplines.

Coupled with the fascination for whatever law enforcement or criminal justice fads are promoted by the entertainment media, such as forensics (for example, *CSI: Crime Scene Investigation*) or law (such as *Law & Order*), the greatest change confronting criminal justice both in the practitioner and academic realms is homeland security.

Notes

1. P. Medhurst, *Global Terrorism* (New York: United Nations Institute for Training and Research, 2002).
2. T.L. Friedman, *Longitudes and Attitudes: The World in the Age of Terrorism* (New York: Anchor Books, 2003).
3. D. Young and J.M. Cooper, Jr., "William McKinley." In M. Beschloss (ed.), *American Heritage Illustrated History of the Presidents*, pp. 305–314 (New York: Crown Publishers, 2000).
4. Buffalo History Works web site, http://www.buffalohistoryworks.com, retrieved March 2, 2008; D. Young and J.M. Cooper, Jr., "William McKinley." In M. Beschloss (ed.), *American Heritage Illustrated History of the Presidents*, pp. 304–313 (New York: Crown Publishers, 2000).
5. G. Caesar, *Incredible Detective: The Biography of William J. Burns* (Englewood Cliffs, NJ: Prentice Hall, 1968); M. Jones, *Criminal Justice Pioneers in U.S. History* (Boston: Allyn & Bacon, 2005).
6. P. Medhurst, *Global Terrorism*.
7. L. Stein, *The Balfour Declaration* (New York: Simon and Schuster, 1961).
8. J.R. White, *Terrorism and Homeland Security*, Fifth Edition (Belmont, CA: Thomson/Wadsworth, 2006).
9. P. Hart, *Mick: The Real Michael Collins* (Viking Press, 2005).
10. R. Alonso, *The IRA and Armed Struggle* (New York: Routledge, 2007).
11. S.K. Aburish, *Nasser: The Last Arab* (New York: St. Martin's Press, 2004).

12. A. Hourani, *A History of the Arab Peoples* (Cambridge, MA: Belknap Press of Harvard University Press, 1991); "Muslim Brotherhood. (2007). In *Encyclopaedia Britannica*. Accessed November 11, 2007, from Encyclopaedia Britannica Online: http://search.eb.com/eb/article-9054456.
13. MIPT Terrorism Knowledge Base, www.tkb.org, retrieved November 10, 2007.
14. M. Jones, *Criminals of the Bible: Twenty-Five Case Studies of Biblical Outlaws* (Grand Haven, MI: Faithwalk Publishing, 2006).
15. P. Medhurst, *Global Terrorism*.
16. J.R. Nassar, *The Palestine Liberation Organization: From Armed Struggle to the Declaration of Independence* (New York: Praeger, 1991).
17. J. Garner, *We Interrupt this Broadcast: The Events that Stopped Our Lives—from the Hindenburg Explosion to the Attacks of September 11* (Naperville, IL: Sourcebooks, 2002).
18. D. Halberstam, *The Fifties* (New York: Villard Books, 1993).
19. F.A. Gerges, *Journey of the Jihadist: Inside Muslim Militancy* (Orlando, FL: Harcourt, 2006).
20. Ibid. p. 178.
21. Bin Laden, Osama (2007). In *Encyclopedia Britannica*. Accessed November 11, 2007, from Encyclopadia Britannica Online: http://search.eb.com/eb/article-9343571
22. G. Crile, *Charlie Wilson's War: The Extraordinary Story of the Largest Covert Operation in History*, p. 522 (New York: Atlantic Monthly Press, 2003).
23. National Commission on Terrorist Attacks upon the United States, *The 9/11 Commission Report: Final Report of the National Commission on Terrorist Attacks upon the United States* (New York: W.W. Norton, 2004).
24. Gerges, *Journey of the Jihadist*, p. 178.
25. National Commission on Terrorist Attacks upon the United States, *The 9/11 Commission Report*.
26. Gerges, *Journey of the Jihadist;* National Commission on Terrorist Attacks Upon the United States, *The 9/11 Commission Report*.
27. National Commission on Terrorist Attacks Upon the United States, *The 9/11 Commission Report*.
28. Ibid.
29. Ibid; Also, the "docudrama" film *United 93* presents an account of this incident.
30. Ibid.
31. Ibid.
32. T.L. Friedman, *Longitudes and Attitudes*.
33. S. Bar, *Warrant for Terror: The Fatwas of Radical Islam and the Duty to Jihad* (Lanham, MD. Rowman and Littlefield, 2006).
34. *Washington Post* Internet Web Site, www.WashingtonPost.com, accessed November 10, 2007.
35. National Commission on Terrorist Attacks upon the United States, *The 9/11 Commission Report*.

Discussion Questions

1. What are the historical events and forces that led to the terrorist attacks of September 11, 2001?
2. What are the implications of labeling efforts to combat terrorism a "war" on terrorism? When can Americans be assured that the end of the "war" is in sight?
3. To what degree is the current war on terrorism a criminal justice effort versus a military effort?

Chapter 3

Globalization, Terrorism, and Homeland Security

By Larry K. Gaines and Victor E. Kappeler

> Nothing is easier than to denounce the evildoer; nothing is more difficult than to understand him.
>
> —Fyodor Dostoevsky

Globalization, terrorism, and homeland security are among the most complex and controversial issues of our day. These issues are complex because they affect almost every aspect of society, and are deeply enmeshed in very dynamic historical, social, cultural, economic, religious, and political contexts. They are controversial because they force us to confront some rather unpleasant facts, reflect on uninterrogated and culturally bounded beliefs, and directly face a world outside our zone of comfort. The issues surrounding globalization, terrorism, and homeland security are further confounded because they involve contradictions that are not easily reconciled, forcing us to confront a host of interrelated social problems and perspectives. The resolution of one problem can have cascading effects on other problems or it can even give rise to new and unanticipated problems. An advance in security can lead to an erosion of civil liberties, greater public awareness about the possibility of terrorism can lead to prejudice and racism, and national security measures can lead to international isolation.

Issues of violent resistance, terrorism, and national security are inherently political, rather than merely rational—they involve questions of power, interests, and perspectives. Making matters worse, terrorism and homeland security are not only inherently political, they have become "politicized" with various groups attempting to advance ideological agendas, rather than rationally confronting the myriad of connected social issues. Immigration policy, for example, is only tangentially related to terrorism but is directly related to globalization. Anti-immigrant groups, however, use issues of security and border protection to advance their ideology of exclusion, while other groups use the fear of exclusionary policies in hopes of reshaping the political and economic landscape or merely benefiting from sources of cheap labor.

In this chapter we address some of the many issues surrounding globalization, terrorism, and homeland security as they affect society and the institution of policing. In doing so, we attempt to present a rational and fact-based view of the issues without avoiding the contextual contradictions, controversies, and often disconcerting aspects of the topic. Rational and pragmatic solutions to the real problems associated with globalization, terrorism, and security can only be found through a candid consideration of the varied contexts from which these problems arise.

Globalization, Conflict, and Policing

The movement toward globalization and the discovery of terrorism as "new" social problems are two interrelated forces that promise to forge the development of American policing well into the twenty-first century. The governmental response to terrorism requires changes in policing, even at the local level. Globalization marks a change in organization of the economic system, which also requires changes in policing. Major changes in the economic conditions of all societies are coupled with transformations in the way social order is produced. The police have consistently adapted to comply with major changes in economic conditions. Shifts in production, the circulation of capital, and the techniques of generating capital all affect the practices of police.

Major Social Revolutions and Police Focus

Economic Engine/Way of Knowing/Geographic Focus	**Police Attention/Economic Orientations**
Pre-modern • Human and Animals • Religion • Local/tribal	**Agricultural Revolution** • Slave Patrols • Morals • Commodities
Modern • Machines • Science • National	**Industrial Revolution** • Strike Breakers • Riots • Contraband
Postmodern • Data and Cyber-Commerce • Signs • Global	**Information Revolution** • Internet • Terrorism • Political Dissent

Figure 3.1

The more rapid and drastic an economic change, the more radical the transformation in a society's police structure. When societies undergo major change in economic arrangements or modes of production, it is referred to as an economic revolution. Scholars generally recognize three major economic revolutions: agricultural, industrial, and technological/information. The information revolution can be better described as the movement toward globalization. While globalization has a long history, it is a predominate theme in contemporary world affairs. Less frequently discussed, however, are the accompanying transformations that occur in policing. When a society undergoes an economic revolution, there also is a tendency to realign policing to better fit the emerging political economy.

Social and economic revolutions are not clearly distinct nor do they have a marked beginning or end. The effects of the agricultural and industrial revolutions still linger and influence both the police and society. Police still address rural and urban concerns, they respond to strikes and labor disputes, and they still use force in response to social problems. There are, however, emerging social changes that promise to further transform the police institution well into the twenty-first century—globalization is one of those changes.

Globalization

Many scholars mark the demise of the industrial revolution by the rise of a service and information economy. Some claim the reduction in manufacturing is being replaced with the provision of services, investment banking, the management and sale of information, and that rise of computer and Internet technologies is the new economy. This assertion, however, fails to recognize that the emerging economic revolution has merely dislodged the location of the production associated with the industrial revolution to developing nations, not replaced it. At the same time, there has been a tremendous growth in international and global trade. Immigrants, both legal and illegal, are performing more work in the United States and more jobs are being outsourced to other countries. The American economy is fast transforming from a production-based economy to one based on the coordination, control, and distribution of world capital. The emergence of computer, information, and Internet technologies has hastened this transformation. Where the agricultural revolution saw capital circulate in local venues, and the industrial revolution experienced capital circulation at the national level, the current revolution is witnessing capital being controlled on a global scale. Globalization is, perhaps, a better term for the new economy, rather than service or information.

Globalization is one of the dominant social themes of the twenty-first century. Not unlike the agricultural or industrial revolution, globalization is a concept that captures the current state of social and economic circumstances. While there are various definitions and ways to think about globalization, the term generally connotes an increasing worldwide integration of social, cultural, and economic systems; especially, markets for goods, services, labor, and capital (Appadurai, 1990). Changes in the economic, social, cultural, and geographic boundaries are facilitated by technologies (Lash & Urry, 1987; 1994). In the economic sphere, it is a rise in the influence of large *multinational corporations* (MNC), an emerging world economy, and more frequent intervention into domestic policies of nations by international institutions like the *International Monetary Fund* (IMF), *World Trade Organization* (WTO), and the *World*

Bank (WB). These international institutions become sites of struggle, often-pitting police against political activists. Collectively, these economic changes are seen as creating a global financial system. In this framework, the United States is fast becoming a controller of capital, rather than a producer of goods and services. By controlling the distribution of capital, the engine of the industrial revolution is being relocated to developing nations. The American economy is more dependent on the generation of new capital through the extraction of profit from foreign, rather than domestic labor and making developing nations debtor states. In essence, the American economy is increasingly based on investment capital rather than the domestic production of goods.

Complicating global security is increased competition for resources, especially energy supplies. Third-world countries such as India and China are increasingly becoming industrialized, supplying the United States and other industrialized countries with cheap goods and sources of labor. These countries now consume more energy, often competing with the United States. As competition for resources and cheap labor increases, it results in more world tension. It also results in substantial underclasses that are discontent, and sometimes resort to violence to better their situation. Moreover, countries that produce scarce resources are able to exert greater influence, especially political and economic demands, on countries that consume a large share of resources. This situation sets nations up in complex relations that create both interdependence and competition—a scenario for conflict.

In the social and cultural spheres, globalization denotes a process by which groups that are socially and geographically distinct become more interconnected. This is facilitated by greater travel, immigration, communication, and the exchange of information. The exportation and importation of goods, the "norming" of cultural beliefs and practices through market logic rather than tradition, and the eradication of social and cultural practices that inhibit globalization all act to reduce cultural diversity. Conflicts between cultural practices that support and those that hinder market expansion and logic become center stage. At the same time, economic norming forces produce overt resistance to globalization. Laborers, protesters, activists, and those concerned with issues of social justice are seen as barriers to creating a risk-free opportunity for economic expansion and control of the new economy.

This increased interconnection often does not spur social and cultural cohesion in a political sense. Indeed, many groups have their way of life and economic livelihood threatened as they are pushed into subordinate roles. They do not want to become "westernized," have their culture eliminated, and attempt to restore their cultural and ethnic heritage. Thus, they do not want to remain a part of other social and cultural groupings of states. This, along with many states' violent responses has spawned conflict in a number of countries. The new countries of Yugoslavia were wrought with terrorism and civil war as ethnic and religious groups fight to become distinct countries or enclaves. States often resort to their own forms of terror to prevent the fragmentation of their political geography. While at the same time, nation states are redrawing their political-economic geographies creating trade zones like the European Union or developing trade agreements like NAFTA. All of these changes create an environment ripe for conflict.

Contemporary conflicts, however, have long histories and most are linked to war, colonialism, occupation, and the dividing of political geographies. Most of the modern-day conflicts in the Middle East can be traced to British and French occupation. For example the Israel-Palestine conflict is directly related to British colonialism, Israel's use of terrorist tactics against British administration and Israeli's occupation of Arab lands. Many Muslims in the Middle East view Israel as an invader who uses terrorist tactics and therefore feel justified in fighting the Israeli occupation of Muslim lands. They also attack United States interests for supporting Israel and for taking oil for what they perceive as an unfair price or engaging in trade relations with less than democratic states that repressed their own populations. Muslims and Hindus attack each other in India for perceived religious reasons, but the seeds of their conflicts, like the Israel-Palestine conflict, are grounded in British colonialism and land disputes. The conflicts in Afghanistan can be traced to the cold-war era and the former Soviet Union's occupation of that country and the United States' training, arming and supporting of oppositional groups. There are numerous conflicts and acts of terrorism and violence throughout the world that, to a large extent, are caused by the lingering effects of colonialism, the installation of monarchies, the quest to secure land and resources, globalization, and the struggle for liberation and domination.

The creation of a "risk-free" environment for this new economy requires not only a transformation in policing, but a historical understanding of the basis of conflict and the prospects for addressing these conflicts with police force. This means that there must be changes in who is defined as dangerous, the dominant ideology must be exported, and laws must be conducive to the economic system. New conduits of capital must be opened and secured, and others must be closed or restricted. More importantly, for policing institutions, globalization requires an ability to project force and control on a global scale. The focus of police institutions is being realigned to fit the emerging economic order, often without an understanding of the historical and cultural forces that contextualized these changes and their attendant conflicts. Police must change their focus to those places where capital now circulates, the areas of conflict over resources, and those people deemed to be threatening. Terrorists, political activists, protesters, and others who resist globalization are quickly becoming the new targets of domestic and international policing.

Terrorism, Politics, and Police

Terrorism is a unifying theme in globalization, but it is also a political designation. This situation arose long before the terrorist attacks on 9/11. The past several directors of the FBI attempted to expand the designation of "terrorist" to groups that seek political, economic, and social change (see, Freeh, May 10, 2001). The very definition of terrorism is problematic. In its most simplistic sense, *terrorism* means the threat or use of violence against civilians for the purpose of achieving a political objective. The use of terrorism is thought to instill shock and fear in such proportions that it creates conditions by which social or political change occurs. The label "terrorist," however, is inherently political. Even nation states disagree on which groups they designate as terrorists.

The concept of terrorism must, however, be distinguished from those of political activist, insurgent, revolutionary, or guerrilla as they are not forms of terrorism. While they seek political change and sometimes use violence to that end, these groups do not necessarily target civilians to achieve their political objectives. In this sense revolutionaries, as well as governments, can become terrorists when

International and Domestic Terrorist Incidents by Group Type and Consequences, 1998–2007

Group Classification	Incidents	Injuries	Fatalities
Anarchist	97	6	0
Anti-Globalization	87	15	7
Communist/Socialist	2,257	4,263	2,163
Environmental	82	0	0
Leftist	109	112	29
Nationalist/Separatist	3,268	17,982	7,189
Racist	25	14	1
Religious	2,284	27,253	12,144
Right-Wing Conservative	96	20	262

Source: Data from MIPT Terrorism Knowledge Base retrieved 8/30/2007; categories modified.

Figure 3.2

they use violence to achieve political objectives, if that violence is directed at civilians. It is important to understand that individuals, groups, or even governments can commit acts of terrorism.

What constitutes terrorism or a terrorist is being redefined by globalization. This reconstruction has become highly politicized and is increasingly being defined in economic terms. For example, the former Director of the FBI attempted to extend the scope of terrorism to groups that are clearly outside the classic definition of terrorism. Louis Freeh made the following remarks in a speech on terrorism well before the attacks of 9/11:

> "Anarchists and extreme socialist groups—many of which, such as the Workers' World Party, Reclaim the Streets, and Carnival Against Capitalism—have an international presence and, at times, also represent a potential threat in the United States. For example, anarchists, operating individually and in groups, caused much of the damage during the 1999 World Trade Organization ministerial meeting in Seattle" (Freeh, May 10, 2001).

It is apparent that what each of these groups have in common is a demonstrated resistance to both free-market capitalism and globalization. Additionally, it is equally apparent than none of the groups mentioned by Former FBI Director Freeh target innocent civilians with violence and almost all have denounced terrorism as a viable tactic for political change. Ironically, none of these named groups have been responsible for a terrorist incident and, as shown in Figure 3.2, even the overly broad label "anarchists" cannot be attached to a single terrorist fatality worldwide.

In similar fashion, law enforcement and political leaders are now using fear and rhetoric to gain unrestricted access to the new methods of transferring and circulating capital in a global economy. Claims made about dangerous illegal aliens, cyber-terrorists, identity theft, child pornography, Internet fraud, and weapons of mass destruction, to some extent have replaced the traditional concerns about street crime and even the "quality of life" and "weed and seed" language of just a decade or so ago. Urban space, street crime, and community policing as focal concerns have been dislodged with talk of

International and Domestic Terrorist Incidents by Region, 1998–2007

Region	Incidents	Injuries	Fatalities
Africa	439	8,127	2,396
East & Central Asia	128	393	164
Eastern Europe	1,362	4,980	1,943
Latin America & the Caribbean	1,785	2,572	1,653
Middle East/Persian Gulf	12,656	46,634	24,522
North America	120	2,408	2,996
South Asia	4,786	17,672	7,469
Southeast Asia & Oceania	1,462	4,952	1,519
Western Europe	3,071	1,780	401
TOTAL	25,809	89,518	43,063

Source: Data and categories from MIPT Terrorism Knowledge Base retrieved 8/30/2007.

Figure 3.3

global terrorism, domestic surveillance, Internet crime, international trade in humans, narco-terrorists, and coordination of international law enforcement activities. All these definitional transformations are designed to shift the focus of the police institution to those who represent resistance to changes in the political economy. Thus, terrorism

> seems to depend on one's point of view. Use of the term implies a moral judgment; and if one party can successfully attach the label terrorist to its opponent, then it has indirectly persuaded others to adopt its moral viewpoint. Hence the decision to call someone or label some organization 'terrorist' becomes almost unavoidably subjective, depending largely on whether one sympathizes with or opposes the person/group/cause concerned. (Hoffman, 2004:260; internal citations omitted)

Defining Terrorism and Its Variations

Today, there are terrorist groups throughout the world that are waging war on countries, institutions, political ideologies, and political or ethnic groups. Recently, we have witnessed major terrorist events in the United States, Spain, Great Britain, Russia, Japan, Kenya, Tanzania, Turkey, Saudi Arabia, and Indonesia, to name just some of the affected countries. Moreover, these events are considerably more deadly, sophisticated, and better planned as compared to those in the past. We are witnessing attacks on high-profile targets, and the attacks are carried out using weapons that inflict substantial death, injury, and damage. The fear of terrorism comes, in part, from the aggregation of incidents and presenting terrorism as affecting the entire world. As Figure 3.3 shows, however, terrorist attacks are not evenly distributed across regions and North America (the United States and Canada) has experienced the fewest incidents of terror. Likewise, even horrific acts of terrorism differ with some attacks requiring months of planning and coordination and the expenditure of hundreds of thousands of dollars. For example, the *9/11 Commission* estimated that the terrorist attacks in Washington and New York cost between $400,000 and $500,000. Other high-profile attacks, like the London transit bombing, are funded with a few thousand dollars gained from credit card fraud and by passing bad checks. Still other incidents receive funding from nation states. While, terrorism may seem to have reached new levels of sophistication and destruction, there is wide variation between incidents.

It is difficult to define terrorism because there are so many attacks, and these attacks have a variety of intertwined motivations including economic, political, social, and religious. Moreover, there are multiple groups committing terrorist acts who have different and sometimes multiple motivations. As such a variety of definitions have evolved but each definition has an inherent political perspective and is designed to define differently those who use violence to maintain versus those who seek to use violence to secure political power.

Under United States Federal Criminal Code, Chapter 113B of Part I of Title 18, terrorism is defined as

> activities that involve violent ... or life-threatening acts ... that are a violation of the criminal laws of the United States or of any State and ... appear to be intended (i) to intimidate or coerce a civilian population; (ii) to influence the policy of a government by intimidation or coercion; or (iii) to affect the conduct of a government by mass destruction, assassination, or kidnapping; and ... occur primarily within the territorial jurisdiction of the United States ... [or] ... occur primarily outside the territorial jurisdiction of the United States.

Although the United Nations has never adopted a formal definition of terrorism, primarily because it would need to include acts of state terrorism, one UN panel described it as an act:

International and Domestic Terrorist Incidents by Tactic, 1998–2007

Tactic	Incidents	Injuries	Fatalities
Armed Attack	6,682	7,765	11,826
Arson	850	179	295
Assassination	1,719	1,122	2,484
Barricade/Hostage	55	1,412	604
Bombing	14,290	75,137	22,710
Hijacking	31	5	26
Kidnapping	1,529	143	1,353
Other	149	427	118
Unconventional Attack	56	3,019	3,047
Unknown	448	309	600
TOTAL	25,809	89,518	43,063

Source: Data and categories from MIPT Terrorism Knowledge Base retrieved 8/30/2007.

Figure 3.4

> intended to cause death or serious bodily harm to civilians or non-combatants with the purpose of intimidating a population or compelling a government or an international organization to do or abstain from doing any act (United Nations Panel, 2007).

Laqueur (1987) notes that terrorism is the "illegitimate" use of force to achieve a political objective by targeting innocent people. The FBI has defined it as, "the unlawful use of force or violence against persons or property to intimidate or coerce a government, the civilian population, or any segment thereof, in furtherance of political or social objectives." These definitions, of course favor those who have the power to define what constitutes "illegitimate" and "lawful" violence or force. Thus, to a great deal, terrorists are those without political or social power who attack the powerful, usually the government and its representatives or corporations, to wrestle power from political elites. Because they are not powerful enough to attack the government itself in conventional terms, they attack civilians and critical infrastructure to create fear, chaos, and economic instability. To the civilian injured in a terrorist attack it makes little difference if a state's covert military personnel, a political activist, or a religious zealot detonates the bomb.

Terrorism represents the lowest level of conflict that can be waged primarily because the combatants often represent a group without state power and they are without the resources to fight on a larger conventional scale. If terrorists are successful, they will create an atmosphere whereby others become interested in their cause. They will also cause economic and social hardships that result in the government taking punitive, drastic courses of action that often are aimed at innocent civilians. Because terrorists can blend in with the population, it is difficult to differentiate terrorists from civilians. When the government responds to terrorist acts with repression or retracting civil liberties, the terrorists often are able to gain more support and recruit more people to join the fight. This can escalate into guerilla warfare where the opposition gains the personnel and resources to directly fight the government. If the government continues to repress the citizenry, the opposition continues to gain support, the guerilla war can evolve into civil war and, with sufficient support, the emergence of a new state. Unfortunately, history shows that nations are often born of terrorism. Although terrorism without the political support of a population cannot result in the overthrow of a government, it can lead to actions and conditions that ultimately accomplish this objective. It is, however, a risky and deadly tactic that most political groups denounce.

Terrorism has its roots in economic inequality and repression, political power, and religion. There are numerous countries where the governments are run by and for the political and economic elite and repress large segments of the population. These elites amass large amounts of wealth while the majority of people are poor, living in substandard conditions, and often without food, shelter, and other basic essentials. Ultimately, small groups rise up and begin terrorist campaigns against the ruling classes. Sometimes they use terrorist or small guerilla operations. Many countries are hotbeds for these types of problems. Political power often involves one political party attempting to wrestle power from a ruling political party. They often have militias and money to fund the conflict. Often this funding comes from nations that have a direct political interest in supporting the groups and overturning an existing regime.

In other situations, nations fund and support other states in their exercise of political repression against their own minority populations fueling conflict. When nation states provide the economic and military resources necessary to repress citizens they may become the targets of terrorist activities.

Religion plays a key role in a substantial amount of terrorism. White (2006) notes that religious violence centers around three causes. First, some religious groups feel they must "purify" the world in the name of their god. For example, some Muslims believe that all Christians and Jews should be killed. Likewise, some Christians have bombed abortion clinics, killing innocent health care workers. Other Christian groups have killed minorities and attacked governments. Second, some groups believe they are the "chosen ones" and they must destroy all other people in the name of righteousness to defend themselves or the lands they have taken from others. In this sense, religion is not a cause of terrorism, it is used as one motivation for terrorism. Finally, people may become so consumed in a particular cause that they resort to violence to advance their beliefs.

One question often asked by the public and raised in the media is why United States interests are the targets of terrorist groups. There are three primary reasons. First, the United States is seen as a principle supporter of many repressive governments. The United States, for example, supported and continued trade relations with the South African government during apartheid. Also, United States support for Israel has resulted in substantial hatred. American support for Israel is not merely political support; Israel is the world's largest recipient of American economic and military aid. When Israeli helicopters attack positions in Palestine, often killing innocent civilians, their military are flying American built and supplied aircraft, a fact known to the Palestinian people. Many Muslims and Christians alike see Israel as occupying Arab lands and repressing Palestinians and other Middle Eastern peoples. The United States supports a military dictatorship in Pakistan, trades with and supports a repressive monarchy in Saudi Arabia, and has overthrown democratically elected governments around the world. For example, many of today's conflicts with Iran stem from American involvement in the overthrown of elected Prime Minister Mossadegh and the instillation of a brutal monarchy, which later resulted in an Islamic revolution. Second, American policies favor transnational corporations, which are seen by many as repressive and failing to use fair and appropriate business practices. One motivation for the American- and British-led overthrow of Iranian leadership was Mossadegh's nationalizing the Anglo-Iranian Oil Company (which later became British Petroleum, BP). Finally, many see the invasion of Iraq as imperialism, spawned by a desire to take Middle Eastern oil. Regardless, there is ample anger toward the United States, making its interests and assets prime targets for terrorists.

Weapons Of Mass Destruction

Although terrorists have never used a *weapon of mass destruction* (WMD) against the United States and most terrorists use conventional or "homemade" weapons, WMDs are of great concern to political leaders, the media, and the public (Henry & King, 2004). WMDs can be biological, chemical, or nuclear. WMDs have the potential capacity to cause large numbers of deaths, injuries, and destruction, especially in heavily populated areas. For example, exploding a small nuclear device in Chicago, New

York, or Los Angeles would inflict substantial destruction. Moreover, it would have long-term residual effects. It would overload first responders and hospitals. It would substantially affect the economy having an impact on thousands of persons who were not directly affected by the explosion. In essence, in addition to the initial loss of life and property, it would cripple a city, resulting in numerous problems that could last for years or decades. To some extent, the primary purpose of using WMDs is not the initial death and injuries. Destroying a primary communications center might result in initial deaths and injuries and loss of property, but the loss of communication would have a dramatic impact on a society. WMDs include a wide range of weapons potentially more dangerous than conventional weapons.

A "WMD" is any explosive, incendiary, poison gas, chemical, biological, or nuclear device that can cause substantial widespread devastation. WMDs differ from conventional weapons because they target a wide, rather than a specific, area. WMDs have the potential to spread destruction across a wide area affecting hundreds, possibly thousands, of people. Moreover, they have long residual effects. For example, it would require considerable effort, time, and expense to clean up an area attacked by a biological, chemical, or nuclear device. These long-term effects would result in great levels of health problems, public fear, and a disruption to the economy.

Nuclear Weapons

Of all the WMDs, nuclear weapons raise the greatest concern. Even a small nuclear device detonated in a large city could result in catastrophic destruction and a large number of casualties. Moreover, the presence of nuclear materials—radiation—would result in long-term problems for a country.

Essentially, a nuclear device can be constructed from highly enriched uranium or plutonium. There are two ways terrorists could acquire a nuclear weapon. First, they could steal or purchase one constructed by a nation state. Second, terrorists could acquire the materials from a nation and construct one themselves. Of these two possibilities, the Union of Concerned Scientists notes, the greater likelihood is that terrorists would attempt to acquire the materials and construct a device themselves. If terrorists are able to acquire all the necessary components, it is not difficult to construct a nuclear weapon. Only a small amount of fissile material is required to build a bomb. A crude weapon could be constructed from about 12 kilograms of highly enriched uranium or four kilograms of plutonium. There are several Middle Eastern countries that currently have the knowledge and capability to build these weapons, and some of these countries are sympathetic to extremist causes. Indeed, scientists in Pakistan have shared technical information with other countries in the Middle East. Moreover, a number of extremist groups have the funding necessary to purchase materials, expertise or weapons.

Another problem is that the materials necessary for building these weapons are stored inadequately, especially in Russia and the countries that broke from the former Soviet Union. In many cases, the materials are stored with substandard security, and there are large quantities unaccounted-for materials. Even the United States cannot account for all of its fissile materials. Additionally, the production of tactical battlefield nuclear weapons by the United States makes these devices smaller and easier to transport. A tactical nuclear weapon can easily fit into a piece of luggage.

In terms of nuclear weapons, three different threat scenarios exist and each has a different level of probability: (1) a dirty bomb, (2) attacks on nuclear power plants, and (3) diversion of fissile material or weapons. In terms of *dirty bombs,* these bombs use conventional explosive materials but are wrapped or contain some type of radioactive material. The radioactive materials are dispersed as a result of the conventional explosion resulting in contamination. A dirty bomb does not necessarily have to contain highly enriched uranium or plutonium; it could contain radioactive waste products that are produced at commercial power plants. Radioactive waste generally has fewer security precautions as compared to highly enriched uranium or plutonium. Additionally, Iraq is littered with hundreds of thousands of radioactive depleted uranium ammunition rounds used by the United States military in the Gulf Wars.

There have been several plots and a single attempt to use dirty bombs. In 1996, Islamic rebels from Chechnya planted a dirty bomb in a park in Moscow. Although not detonated, it contained dynamite and Cesium-137, a by-product of nuclear fission. In 2002, Jose Padilla a Brooklyn-born convert to Islam, also known as Abdullah Al Muhajir, was arrested by federal authorities for plotting to construct and detonate a dirty bomb in the United States. He was held in military detention for almost four years without the benefit of counsel until the Justice Department faced a Supreme Court showdown over his detention and transferred him to the criminal justice system for trial in federal court. In 2007, Padilla was convicted; not of plotting to use a dirty bomb as was initially claimed, but rather of conspiring to kill people and support terrorism—all overseas. Prosecutors failed to present any evidence of the "dirty bomb" allegation to the jury.

The effectiveness of dirty bombs as WMDs, however, is questionable. Because a dirty bomb has never been detonated in public, it is difficult to predict the impact. Research generally shows that these devices are not very efficient as WMDs for two reasons. First, to inflict massive damage, these devices would need to be very large in terms of both the amount of radiation they would have to release and in terms of the amount of conventional explosives necessary to disperse the radiation. This makes dirty bombs impracticable for their destructive features. Second, research predicts these devices would neither sicken nor kill a substantial number of people as compared to other WMDs. The use of one of these devices, however, would instill great panic and potentially disrupt the economy. While the media has focused public attention on these devices, there are far more destructive devices and techniques that should be of concern.

Crashing a large aircraft or large amounts of explosives at a *nuclear power plant* could have disastrous effects as compared to dirty bombs. Such an explosion could cause a reactor's core to melt down (such as occurred at the Chernobyl nuclear power plant, located in the former Soviet Union) or cause spent fuel waste to be spread across a large geographical area. The effects could be devastating, and the clean up could take decades. Russia is still dealing with the aftermath of the Chernobyl disaster, which occurred in 1986.

Finally, diversion of *nuclear material* remains a threat. There are numerous sites throughout the world that contain the materials needed to make a bomb, and in many instances, the security is suspect. Moreover, there are countries that are sympathetic to terrorist causes or need hard currency, and they may be tempted to provide terrorists with a bomb or bomb-making materials. Examples include states from the old Soviet Union, North Korea, and Pakistan. The actual threat of a nuclear attack by terrorists, however, must be put into perspective. FBI Director Robert S. Mueller (2007) remarked that, "Although a nuclear terrorist attack is the least likely to occur due to the required technical expertise and challenges associated with acquiring weapons-usable material, the intent of terrorists to obtain this material is a continuing concern."

Biological Weapons

Although given far less attention by the media, *biological weapons* represent a more realistic threat to the United States and the rest of the world. There are a number of contagions that could pose an enormous health threat to large numbers of people. Given diseases such as small pox and anthrax, there is the potential to infect large numbers of people over a wide geographical area. Containment and prevention are perhaps the greatest issues. There is a great deal of speculation that terrorist groups possess or are attempting to possess biological weapons. There was substantial publicity and public fear in 2001 when someone sent anthrax via mail to several locations in the United States causing five deaths and infecting at least 17 people. While the case remains unsolved, some believe that right-wing domestic extremists in the United States carried out the attacks. Evidence shows that bacterial strain used in the attacks was derived from the Ames strain first studied at the *U.S. Army Medical Research Institute of Infectious Diseases* in Fort Detrick, Maryland and later transferred to research labs across the country.

There is some confusion over what exactly constitutes bioterrorism. Here, *bioterrorism* means the use of microorganisms to inflict harm on a civilian population. This definition includes terrorist groups and other groups such as organized crime or hate groups. It should be remembered that biological toxins are used for a variety of rationales. Moreover, the target for a biological attack may not be humans; toxins have been used to destroy crops and livestock. This exemplifies how biological warfare can have multiple targets and devastating effects on living conditions and local economies. The *United States Centers for Disease Control and Prevention* (2007:1) defines a bioterrorist attack as,

Chemical Weapons of Mass Destruction

According to the CDC, the basic categories are:

- Biotoxins
- Blister Agents/Vesicants
- Blood Agents
- Caustics (Acids)
- Choking/Lung/Pulmonary Agents
- Incapacitating Agents
- Long-Acting Anticoagulants
- Metals
- Nerve Agents
- Organic Solvents
- Riot Control Agents/Tear Gas
- Toxic Alcohols
- Vomiting Agents

There are a variety of toxic compounds contained within each of these categories.

Figure 3.5

> "the deliberate release of viruses, bacteria, or other germs (agents) used to cause illness or death in people, animals, or plants. These agents are typically found in nature, but it is possible that they could be changed to increase their ability to cause disease, make them resistant to current medicines, or to increase their ability to be spread into the environment. Biological agents can be spread through the air, through water, or in food. Terrorists may use biological agents because they can be extremely difficult to detect and do not cause illness for several hours to several days. Some bioterrorism agents, like the smallpox virus, can be spread from person to person and some, like anthrax, cannot."

As Ackerman and Moran (n.d.) note, there are three types of biological agents suitable for bioterrorism.

1. Bacterial organisms such as those that cause anthrax, plague, and tularemia.
2. Viruses including those that cause smallpox and ebola.
3. Toxins including botulinum toxin (derived from a bacterium), ricin (derived from the caster bean plant), and saxitoxin (derived from marine animals). Toxins are not alive and cannot multiply like bacteria or viruses, and therefore have the same attributes associated with chemical weapons.

Biological agents are not biological weapons. Mere possession of an agent does not make it a weapon. The agent must be *weaponized,* that is, a would-be terrorist must develop or possess a mechanism that can disperse or disseminate the agent. There is an incubation period associated with most biological agents. That is, there is a time period (one to two weeks) before there is an onset of the disease within the host or victim. This allows perpetrators to distance themselves from the scene before law enforcement officials become aware of the act. It also provides a time for medical experts to begin responding to the problem. Also, contagious and noncontagious biological weapons exist. For example, anthrax is not contagious like smallpox. We know very little about the effectiveness of biological weapons in terms of casualties. Most of the information we have is from military uses and experiments. However, perhaps the greatest problem with many biological agents is the effect on the population. An outbreak of smallpox could result in a great deal of disorder, because it is a highly contagious virus, transmits rather easily, and results in a very high mortality rate. Concerns are even greater because the virus is still stored in Russian and American laboratories.

As for the possibility of a terrorist group using these weapons against the United States, FBI Director Robert S. Mueller (2007) testified before Congress that, "Few if any terrorist groups are likely to have the capability to produce complex biological or chemical agents needed for a mass casualty attack, but their capability will improve as they pursue enhancing their scientific knowledge base by recruiting scientists as some groups are doing."

There are several reasons why a group or country might use bioterrorism. There are strategic motivational factors. For example, the use of bio-weapons could result in a specific objective: (1) mass casualties, (2) economic damage, (3) target contamination, (4) disruption of the enemy's ability to conduct business as normal, and (5) undermine public confidence in the target government. When used strategically, such weapons are seen as one tactic that is coupled with others. At the same time, there are constraining factors including the perception that these weapons may be counterproductive—reprisals and alienation of support groups, and that it may be opposed by state-sponsors.

Chemical Weapons

As displayed in Figure 3.5, the *Centers for Disease Control and Prevention* (CDC) has compiled a listing of the various chemical compounds that may be used in some type of chemical attack.

Chemical warfare was a primary weapon used in World War I and in the Iraq-Iran War. One million casualties occurred in World War I as a result of combatants using chemical weapons. Iraq used chemical weapons almost daily against Iran in that conflict. Among the largest manufactures of chemical weapons are firms in the United States, Germany, Britain, and France. Thus, chemical weapons are not new; nor are they produced in any meaningful quantity by terrorists. Ironically, countries continue to develop new and more deadly compounds that pose a risk to themselves should they be used by terrorists.

First, different *chemical weapons* have different levels of toxicity and lethality. Some of the chemicals used are insecticides, which have a low level of toxicity, but are readily available. Moreover, while there is evidence that insecticides have been used the results have been limited to nonexistent. On the other hand, weapons grade chemicals are potentially dangerous. Chemical weapons like mustard gas, Sarin, and VX are extremely lethal. As with biological weapons, the effectiveness of a chemical threat is largely based on the delivery system. Many experts believe that chemical weapons cannot produce the same level of casualties as nuclear or biological weapons. To have a significant impact, hundreds of thousands of pounds of the chemical would have to be effectively released. Although governments have used chemical weapons for decades, one of the very first terrorist uses of chemical agents against civilian population did not occur until 1995. The attack was carried out in Japan. Aum Shinrikyo, an apocalyptic cult group released Sarin into the Tokyo subway. The incident killed at least 12 people and injured more than 5,000 people. In essence, an effective attack would only be capable of causing hundreds to a few thousand casualties. A primary difference between chemical agents and biological agents is that the chemical agents have an immediate effect. This results in crippling any response or providing any warning to civilian populations.

Because the vast majority of all WMD and their necessary delivery systems are produced, distributed, sold, stored and used by nation states, a logical and rational first step in combating terrorism would be to either eliminate or restrict the distribution and sale, or to adequately store and secure the materials before they become weapons of extremist groups.

Extremist Groups

There are literally hundreds of extremist groups around the world. These groups can work in isolation, they can be connected to a main group or cause, they can be "sponsored" by states, or they can be state-sanctioned groups. As with the concept of terrorism, it is difficult to define exactly what constitutes an extremist group. We might initially think of *extremists* as those groups that constitute a terrorist organization, have a violent political orientation, or posses a militant orientation to an issue. We must also keep in mind that extremists come from almost all countries, races, religions, and ethnic groups. There are white Christian extremist groups living in the United States like the Christian Identity Movement, Christian Defense League and the Covenant, The Sword, and the Arm of the Lord; Jewish extremists who live in Israel like Gush Emunim, Jewish Fighting Organization and the Kach Party as well as Arab Muslim extremist groups. In the sections that follow we briefly consider a few of the many extremist groups to illustrate the point.

Extremist Groups and the Middle East

It is only in the after math of 9/11 that the American media and politicians have turned their focus on Arab Muslim extremists. Despite this current unprecedented attention, very few Muslim groups have expressed a desire to do domestic harm to the United States (NIC, 2007). It should be remembered that hatred for the United States rests with a large number of political conflicts that have spanned both history and geography. One of these major conflicts is the Israeli and Palestinian conflict, which has been waged for more than one-half century. This conflict, with its bombings, wars, fractionalization, and sustained resistance provides fuel to the Middle East. The scope and long timeline of the conflict has resulted in a number of parties becoming embroiled in the conflict. This is not to say that other conflicts or interests have not resulted in anger toward the United States. There are groups throughout the rest of the world who also have directed their anger toward the United States because of it policies, but those numerous groups in the Middle East seem the most outraged. For example, President Ronald Reagan called the Afghan Mujahideen "freedom fighters," when they were at war with the former Soviet Union over its occupation of their country. After the fall of the Soviet Union, an armed student movement arose in Afghanistan, with the backing of Pakistan and the United States—the Taliban. After the 9/11 attacks, the United States ousted the Taliban and formed a new government under Hamid Karzai. Today, President George W. Bush has labeled some of the very same people "terrorists" when they use violence to fight foreign occupation of their country. Policies and conflicts of this nature fuel anger and resentment against the United States.

Hamas

Hamas (Islamic Resistance Movement) is one of the larger extremist groups operating in the Middle East. It is housed in Palestine, and it is at the center of the Israeli-Palestinian conflict. After recent elections, Hamas won a controlling number of seats in the Palestinian Parliament. Relations with Fatah deteriorated after it refused to surrender control of the Gaza Strip to the elected government. Hamas was ousted from the government and replaced with members of Fatah. Currently Hamas controls the Gaza Strip and Fatah is in control of the West Bank with the two groups fighting for ultimate control of Palestine. Many of its members are well-educated, possessing advanced degrees in science and engineering. The group is made up of two wings: a political wing that provide social services for the Palestinian people like hospitals, schools and libraries, and a militant wing (Izz ad-Din al-Qassam Brigades) that carries out acts of resistance to the Israeli occupation. Many of Hamas' activities are funded by Saudi Arabia, Kuwait, and Iran. Hamas enjoys great support among Palestinians because of corruption and ineffectiveness in previous Palestinian governments and because it is seen as a movement that defends Palestinians from brutal Israeli occupation.

Its militant wing carries out attacks against Israel, including suicide bombings, and often targets civilians. In 2006, the militant wing called on Muslims around the world to attack American targets. The political wing, however, has denounced this statement and claims to be interested only in operations against Israeli occupation. The militant wing is organized into cells with some cells having only a few members. To some extent, leaders may be unable to control the decisionmaking and actions of some of the cells. Although Hamas has never carried out a suicide attack outside Israel, the Gaza Strip, or the West Bank, U.S. officials are concerned because they claim there are covert cells in America.

Hezbollah

Hezbollah (Party of God) is a predominately Shiite organization and is strongest in Lebanon. Hezbollah was first formed during the Lebanese Civil War with the principal goal of eradicating Western colonialism and establishing an Islamic government in Lebanon. Some accuse Hezbollah of carrying out the 1983 Beirut barracks bombing which killed more than 300 American and French troops; the group has, however, denied this claim. While the United States and Britain regard Hezbollah as a terrorist group, most in the Arab world view it as a legitimate resistance force. Like Hamas, Hezbollah enjoys a great deal of popular support among the Lebanese people winning numerous seats in the Parliament. Hezbollah enjoys public support because of its position in the conflicts and because it provides a great deal of social service to the people. Hezbollah gained regional and international recognition when it battled Israel in the Lebanese Civil War (1982–1990) and again during the Israeli invasion of Lebanon in 2006. The organization has ties to Syria and Iran, with Iran providing most of the financial and material support for its operations.

In some of the initial conflicts, Hezbollah used suicide bombings and assassinations to expel Israel from Lebanon but, since that time, it has evolved into a military force that uses conventional weapons. At the same time, Hezbollah has also been the victim of attacks and assassinations carried out by Israel. In its movement away from suicide bombings and random attacks Hezbollah began capturing Israeli soldiers to force prisoner exchanges. One of these captures sparked the 2006 Lebanon War. As demonstrated in the 2006 invasion of Lebanon by Israel, Hezbollah can field a substantial fighting force, and possesses a large arsenal of weapons including mortars and rockets. Hezbollah was able to declare victory because the Israeli army could not eliminate them and they inflicted unprecedented causalities on Israeli forces ultimately driving them from Lebanon. Moreover, afterwards, many groups who traditionally had not supported Hezbollah supported the group because of Israel's indiscriminate attacks on civilian targets and the rebuilding programs implemented by Hezbollah after the conflict. The NIC (2007:6) notes that while Hezbollah "has conducted anti-U.S. attacks outside the United States in the past, [it] may be more likely to consider attacking the Homeland over the next three years if it perceives the United States as posing a direct threat to the group or Iran."

Islamic Jihad

Islamic Jihad is a label used by several groups in the Middle East including the Islamic Jihad Movement in Palestine, the Egyptian Islamic Jihad, the Turkish Islamic Jihad and the Yemeni Islamic Jihad. It is claimed that these groups of loosely connected cells have carried out attacks across the Middle East. The Islamic Jihad Movement in Palestine was formed in the Gaza Strip during the 1970s and grew out of the Egyptian Islamic Jihad. The Palestinian Islamic Jihad is a highly secretive offshoot of the Muslim Brotherhood, a Sunni group, that desires social reforms based on Islamic principles. It is smaller than either Hamas or Hezbollah and lacks the resources to provide social services. Like other organizations, the Palestinian Islamic Jihad is composed of both political and armed wings. Its militant wing, the Al-Quds brigades, has been involved in many attacks against Israel, including suicide bombings. The movement was given a setback in 1995 when Israeli Mossad agents reportedly gunned down its leader, Fathi Shaqaqi, in Malta. The group's reach and activities were further limited by a number of arrests and indictments following the 9/11 attacks. Beginning in 2000, the group became more active claiming responsibility for a number of attacks and showing greater coordination with other Palestinian militant groups.

Al Qaeda

Al Qaeda (the base) is perhaps the most publicized and feared terrorist group in the world. The Sunni Muslim group is led by Osama bin Laden and might be better thought of as a network of groups rather than a single formal organization. The group's origin can be traced to Dr. Abdullah Azzam, who was a leader in the anti-Soviet jihad after the Soviet invasion of Afghanistan. Bin Laden became involved in the resistance to Soviet occupation recruiting, funding and organizing the mujahideen (Afghan Arab), which was supported covertly by the United States.

Al Qaeda's espouses an ideology of "defensive jihad" which responds to what is sees as attacks against Muslims, the occupation of Muslim lands, and the operation of repressive governments like Saudi Arabia, Egypt, Algeria, and Afghanistan. The group and its network seeks to expel Western forces from "holy territories" like the Gulf and Iraq. Al Qaeda has shifted its base from Afghanistan to Sudan and back again to Afghanistan and Pakistan depending on strategic considerations and current American war effort.

Al Qaeda, is often claimed, has cells located throughout the world which operated in as many as 65 countries, has a membership in the tens of thousands, and was responsible for the 9/11 World Trade Center attack. These claims, however, are difficult to substantiate since the very nature of al Qaeda's networked structure makes it hard to accurately estimate its size, its global reach, or even if its members are directly responsible for particular attacks. Further complicating the matter is evidence that the 9/11 operation was financed by nation states, conduced by hijackers from Saudi Arabia, United Arab Emirates, Lebanon, or Egypt. A 2007 *National Intelligence Council* (NIC, 2007:6) report stated that, "Although we have discovered only a handful of individuals in the United States with ties to al-Qa'ida senior leadership since 9/11, we judge that al-Qa'ida will intensify its efforts to put operatives here."

Al Qaeda is thought to be responsible for the 1998 U.S. embassy bombings in Africa and the 2000 attack on the USS Cole in Yemen. As with many extremist groups, it is often difficult to determine actual facts from the exaggerations of these groups. Groups will often claim responsibility for sensational attacks to recruit followers to their cause and create a sense that the group is powerful. Additionally, it becomes difficult to distinguish which acts are actually carried out by members of the group itself, want-to-be members, copycats, or individuals who have sympathy for its cause. As researchers McCloud and Dolnik (2002) remark,

> The United States and its allies in the war on terrorism must defuse the widespread image of al Qaeda as a ubiquitous, super-organized terror network and call it as it is: a loose collection of groups and individuals that doesn't even refer to itself as "al Qaeda." Most of the affiliated groups have distinct goals within their own countries or regions, and pose little direct threat to the United States. Washington must also be careful not to imply that any attack anywhere is by definition, or likely, the work of al Qaeda.

Al Qaeda has provided material support and training for a number of other terrorist and insurgent groups. Al Qaeda has trained or supported resistance and insurgent campaigns in Afghanistan, Tajikistan, Bosnia-Herzegovina, Kashmir, Mindanao, Chechnya, Lebanon, Nagorno-Karabakh, Algeria and Egypt. Al Qaeda provides training in camps and sends fighters to support Muslim struggles in other counties. Al Qaeda is part of the fight against American troops in Afghanistan. Some believe that Osama bin Laden and his group have attempted to procure WMDs. Al Qaeda is often alleged to be the organizer of potential attacks by other terrorist groups against the United States, however, the NIC (2007:6) notes that, "al-Qa'ida will probably seek to leverage the contacts and capabilities of al-Qa'ida in Iraq (AQI), its most visible and capable affiliate and the only one known to have expressed a desire to attack the Homeland."

U.S. Locations with Growing KKK Chapters and Membership

SOUTH	MIDWEST	GREAT PLAINS	MID-ATLANTIC
Alabama	Indiana	Iowa	Maryland
Arkansas	Kentucky	Nebraska	New Jersey
Florida	Michigan		Pennsylvania
Georgia	Ohio		West Virginia
Louisiana			
Mississippi			
South Carolina			
Tennessee			
Texas			

Source: Anti-Defamation League (2007). The Ku Klux Klan Rebounds. Washington, DC: Author's Website.

Figure 3.6

Extremist Groups and the United States

We must also consider that there are homegrown extremist groups willing to resort to violence or terrorism.The prime example is the bombing of the Alfred P. Murrah Federal Building at Oklahoma City. On April 19, 1995, Timothy McVeigh and Terry Nichols detonated about 5,000 pounds of explosives killing 168 people. The bombers belonged to a right-wing group who were opposed to the government. The attack was provoked and justified by the terrorists in retaliation for the government's attacks in Waco, Texas and Ruby Ridge. McVeigh said he was motivated by these incidents, a hatred of the government he developed while serving in the Army and by the influence of reading *The Turner Diaries,* a right-wing hate book popular with some racist groups. *The Turner Diaries* is a novel written by William Luther Pierce, a former leader of the National Alliance, a white separatist group that advocated the extermination of all Jews and people of color, to create a "White world." Terry Nichols was tried in 1997 and found guilty of conspiring to build a "weapon of mass destruction" and on several counts of manslaughter. McVeigh was found guilty on several counts of murder, and sentenced to death, and executed in 2001.

There are hundreds of right-wing extremists in the United States with members who might plot a similar incident. As FBI Director Robert S. Mueller (2007), cautioned, "While much of the national attention is focused on the substantial threat posed by international terrorists to the homeland, we must also contend with an ongoing threat posed by domestic terrorists ... Despite the fragmentation of white supremacist groups resulting from the deaths or the arrests of prominent leaders, violence from this element remains an ongoing threat to government targets, Jewish individuals and establishments, and non-white ethnic groups."

The Arizona Patriots

The Arizona Patriots are a group of white supremacists that believe in Posse Comitatus and Christian Identity ideology. In the 1980s, they began planning attacks against the United States government and Jewish Centers. At one point the group sent a threatening letter to state officials in Arizona, demanding their resignation. As a result of undercover operations, several of its members were arrested and convicted of plotting to bomb the Simon Wiesenthal Center, an FBI and IRS office, and the Jewish Defense League. The group also intended to carry out robberies to finance its cause. These plans included the use of homemade bombs, chemical gases, and a plot to destroy the Hoover Dam "as a distraction." Like Timothy McVeigh, these plots were inspired by *The Turner Diaries,* racism, and Christian Identity ideology. Some believe that the group is reconstructing itself, following the imprisonment of some of its members.

Patriot Council

The Patriot Council is an anti-government right-wing group founded by United State Air Force Colonel Frank Nelson. The group draws from radical right Christian Identity ideology and focuses on issues of taxation and governmental power, and rejects the legitimacy of government authority. Where there are no documented cases of terrorism directly link to the group, members of the group considered bombings and the assassination of federal officials. One plot included plans to use Ricin, a biological toxin, against government officials and insulations. Members of the group went so far as to secure the castor beans necessary to make the toxin and actually extracted Ricin from the beans. As a result, members of the groups were arrested and among the first to be charged under the *Biological Weapons Anti-Terrorism Act.* As with most terrorist groups it is difficult to know whether the plot was merely the result of a few rogue members or sanctioned by the group.

Oklahoma Constitutional Militia

The Oklahoma Constitutional Militia, also known as the Universal Church of God, is a small, right-wing, anti-Semitic group. Willie Ray Lampley formed its militant arm in the 1990s. This group networked with other racist Christian Identity groups that advocated American white supremacy. Lampley saw himself as a prophet whose duty was to follow God's edicts to kill "sinful" people like homosexuals, abortion clinic doctors, and civil rights proponents. In the 1990s, members of the group were arrested and convicted for planning to bomb gay and lesbian bars, several abortion clinics, and the offices of the *Southern Poverty Law Center* and the *Anti-Defamation League.*

Ku Klux Klan

The Ku Klux Klan (KKK) is one of the oldest and most well-known terrorist groups in America. The organization rose out of reconstruction in 1866, was founded by veterans of the Confederate Army, and continues to preach hatred against minorities, Jews, homosexuals, Catholics, immigrants, and Communists. The group's ideology is based on a fusing of white supremacy, Christian Identity, and belief in a worldwide Jewish conspiracy. The group was so powerful and intertwined with the state in the 1860s that Congress passed legislation (the Ku Klux Klan Act) to attempt to control the group's activities and control its influence over the justice system.

The group reemerged in the early 1900s and, at the height of its power, claimed somewhere between 4 and 5 million members with local chapters across the country. The group carried out literally hundreds of lynchings and acts of terror and intimidation, from burning crosses, to beating minorities. During the 1950s and 1960s the group murdered women and children bombed churches. One of the most infamous attacks conducted by the KKK was the bombing of the 16th Street Baptist Church in Birmingham, Alabama. The attack took place on Sunday, September 15, 1963 and was designed to instill fear in members of the civil rights movement. KKK members Bobby Frank Cherry and Robert Edward Chambliss detonated a bomb as 26 children were assembling for prayer service. The explosion killed four young teenage girls and injured more than 20 children.

Although the KKK is most often depicted as composed of men from the south that exclusively target Africanas for violence, this was often not the case. The KKK has been active in New England, Canada, the South and Midwest, and has targeted many groups because of race, religion, and political affiliation. The KKK targeted both Communists and Socialists and even killed uniformed Africana soldiers returning home from World War I.

While some argue the group is no longer in existence, several variants of the KKK have emerged at the national level in American society—Imperial Klans of America, American Knights of the Ku Klux Klan, and Knights of the White Kamelia. In 2006, the Anti-Defamation League said, "The Ku Klux

Klan, which just a few years ago seemed static or even moribund … has experienced a surprising and troubling resurgence due to the successful exploitation of hot-button issues including immigration, gay marriage and urban crime." Today, this extremist group is estimated to have at least 150 chapters, and as many as 8,000 members across the nation.

Extremist Groups and Israel

The Jewish Defense League

In 1968, the Jewish Defense League (JDL) was founded in New York City. The group was founded by Rabbi Meir Kahane who later moved to Israel and formed the Kach party. The organization began as a group of vigilantes whose purpose was to protect orthodox Jews. Going far beyond merely protecting New York's Jewish population, the group began to conduct an extensive campaign of terror. Initially the group targeted governments that it thought were anti-Semitic, but rapidly expanded their attacks to include bombings and assassinations. JDL was responsible for the assassination of Alex Odeh, director of the *Arab-American Anti-Discrimination Committee*. The group targeted both Soviet and Palestinian institutions. For decades the group was one of the most violent and active terrorist organizations operating in the United States. In 2001, two JDL members were arrested for planning to kill a Lebanese-American Congressman and blow up a mosque in California. The JDL and its offshoots have a presence in Eastern Europe, Canada, Australia, South Africa, Russia, France, and Britain. In 2005, a dispute over leadership split the groups into factions—JDL, B'nai Elim and the Jewish Task Force. As late as 2007, JDL members were active in political protests and in one case accused of assaulting a controversial Canadian extremist and police at a demonstration (Mahoney, 2007). Ironically, the group is not designated by the U.S. State Department as a terrorist group nor is it on its "watch" list.

Kach

Kach ("Thus") is a political offshoot of the JDL. Kach is an extreme nationalist group that operates in Israel, the West Bank, and the Gaza strip. The group follows what is called "Kahanism" (named after its founder) ideology, which espouses extreme right-wing *Religious Zionist* ideals. These ideas include a belief that Israel should be a theocratic government and that it should reserve citizenship exclusively for Jewish people. To achieve these goals proponents of the ideology would deport all gentiles or reduce their status to resident aliens without political rights.

JDL founder Rabbi Meir Kahane established the group in Israel during the 1970s. The group advocated the annexing of all Palestinian lands and the violent removal of all Arabs from these lands. Before his assignation, Kahane openly advocated the use of terror against both the Israeli government and Arab peoples. In 1984, Kahane was elected to the Israeli Knesset. His movement was opposed to any peace settlement with Palestine. Following the Oslo peace accords, one member of his group killed 29 Muslim worshippers and injured another 125 in a mosque in the southern West Bank city of Hebron. In large part because of the attack, Israel was forced to outlaw the group.

In 2005, the group threatened Israel officials who supported "disengagement" from Gaza. Just before Prime Minister Ariel Sharon's began his plans for disengagement, Eden Natan-Zada, a Kach sympathizer and deserter from the Israeli army carried out an attack against a bus of Arabs in Shfaram, Israel. Today, the group has splintered into many factions, but former affiliates are still active in both violence and running for political office.

Fighting Jewish Organization

The Fighting Jewish Organization (EYAL) was a nationalist group that can be traced to conflicts over British-mandated Palestine and the birth of the State of Israel. The British Mandate was a form of governance in the Middle East between 1920 to 1948. It involved the control of Jordan, Israel, the West Bank, and the Gaza Strip. EYAL members can trace their roots to two right-wing terrorist organizations that carried out terrorist attacks in British Mandated Palestine—the Irgun and the Stern Gang. The group also is an extension of the Kach movement and its ideology. The group is also noteworthy because it shows the historical complexity and controversial nature of terrorism in Israel. As part of the Oslo Peace Accords, Israeli Prime Minister Yitzhak Rabin agreed to turn over the West Bank to the Palestinians. An extremist, Yigal Amir, with ties to EYAL's founder, Avishai Raviv, assassinated Rabin. It was later learned that Raviv was a member of the Israel Secret Police (Shin Bet) investigating right-wing radical groups. Some have speculated that Shin Bet founded EYAL; the group was disbanded after Raviv's exposure. Members of the groups have been absorbed into groups associated with the Kach movement.

Homeland Security

Although *Homeland Security* encompasses all levels of government, it is primarily a federal responsibility. *The Department of Homeland Security* (DHS) is the Cabinet level department with this responsibility. President George W. Bush organized the DHS by combining a number of agencies from other Cabinet level departments. The Department currently has approximately 180,000 employees and is the largest Cabinet level department in the federal government. It was formed to better coordinate the numerous agencies that can play a role in case of a terrorist attack on the United States. Additionally, it has a number of corollary responsibilities. First, through the *Federal Emergency Management Agency* (FEMA) and other agencies within the Department, the DHS responds to natural and manmade disasters including potential terrorist attacks. Second, many people think of terrorism as an attack, but the department plays a role in prevention. There are numerous activities that can potentially prevent an attack or minimize its devastation. Prevention is an important part of security. Additionally, traditional and organized crime are associated with terrorist organizations especially in the areas of procuring financial resources and money laundering. These agencies are involved in thwarting and investigating these activities and crimes, especially as they relate to terrorist organizations.

A major difficulty with the initial organization of the DHS was that agencies transferred into the department had previously established goals, responsibilities, and operational procedures. The move to the DHS resulted in the assignment of new responsibilities and priorities. The difficulties occurred when agencies attempted to integrate new security responsibilities into their current operations (see

Perrow, 2006). Horizontal integration issues made it unclear as to what the agencies' new responsibilities were, and no organizational structure existed to ensure that new responsibilities were conducted effectively. Integration of new responsibilities into the agencies has been piecemeal and evolutionary.

There were also issues of vertical integration. The assignment of new tasks and responsibilities clouded the relationships among the various agencies. Successful completion of new tasks, required innovative working relationships and coordination with other agencies, and these relationships and lines of communications had to be created. Bureaucratic entanglements slowed and sometimes prevented these relationships from developing effectively (Perrow, 2006). Perhaps the best example of this problem was the sharing of intelligence among agencies. In many cases, when agencies were able to gather intelligence information they failed to share it with other agencies, especially agencies that needed the information. This is exemplified by the creation of an intelligence czar, U.S. Director of Intelligence. This office was created to coordinate intelligence and ensure that agencies shared information, but the office's effectiveness remains suspect. Wise (2006) notes that the Department has focused on horizontal integration, which is ineffective given the size and disparity of organizational sub-units. He advocates that the organization should focus on networks and the integration of processes as opposed to hierarchal control. An emphasis on control often stifles innovation and results in greater suboptimization, two bureaucratic consequences that can substantially impair efforts to counteract terrorism. For example, "Immigration and Customs Enforcement agents are dropping terrorism investigations or steering them in other directions because they don't want to work with FBI agents, according to a joint report by the inspectors general of the Department of Homeland Security and the Department of Justice" (Stables, 2007:1).

Compounding the problem are the working relations with state and local governments. To a large extent, prevention and mitigation have become responsibilities of state and local officials even though there are several federal agencies such as FEMA involved in these activities. There have been too few efforts to systematically involve state and federal agencies in strategic and operational planning. Although there have been enhancements in "first responses" through grants and training, little has been done about threat identification, threat assessment, and target assessment.

Most efforts to assist state and local units of government, including the police, have hinged on block grants to the states with little coordinated direction. Block grants allow the federal government to give the appearance of action, but it does not have to assume any responsibility, should the monies be spent in areas that are not productive. This has resulted in a patchwork of action. The federal government, however, has failed to fully fund a number of the required initiatives, placing greater strain on state and local governments. O'Hanlon (2003) notes that funding has been inadequate for those cities most likely to be targets. For example, Hobijn (2002) projected that in 2002, homeland security cost the states approximately $4 billion and the U.S. Conference of Mayors projected the costs to be $2.1 billion for larger cities. The security initiative has resulted in a number of unfunded liabilities.

The DHS's primary responsibility is prevention. That is the DHS is seen as the agency coordinating all efforts to prevent attacks on the United States. As such, the DHS devised the National Infrastructure Protection Plan to provide prevention guidance.

The National Infrastructure Protection Plan

In 2006, the DHS published the *National Infrastructure Protection Plan* (NIPP). This document is a guideline for organizing security in the United States. The plan is designed to provide coordination among the numerous federal, state, and local agencies in protecting the nation's critical infrastructure and key resources. Essentially, the plan enumerates a process consisting of six steps that are central to protecting physical, cyber, and human assets:

1. Establish security goals or performance targets that constitute a protective posture.
2. Identify assets, systems, networks, and functions within and out side the United States that require a level of protection.
3. Assess risks in terms of a direct or indirect attack on particular assets and the probability that a target will be attacked.
4. Establish priorities in terms of risk and levels of current protection and mitigation systems.
5. Implement protective programs for those highpriority assets, especially those that currently have low levels of protection.
6. Measure effectiveness in terms of progress toward hardening assets and preventing attacks (DHS, 2006:30).

Woodbury (2005:2) provides a simpler prevention model especially as it applies to critical infrastructure. He notes that the process includes four sub processes:

1. Threat identification—identify those groups and individuals who would commit terrorist attacks and monitor those groups to determine immediate and long-term intentions.
2. Threat assessment—once potential adversaries and intentions or plans are identified, determine their capabilities to conduct an attack. In other words, do they have the financing, logistical support, and personnel?
3. Target assessment—is the identification of possible and plausible targets. What critical infrastructure and population concentrations are likely targets and what degree of protection currently exists and what actions should be taken to harden targets? What resources are currently available to prevent and mitigate a possible attack?
4. Response-protection decisions—is the development of a process to make decisions about risk management.

These four sub-processes comprise a national prevention system, and as such, each of the four sub systems must be addressed for adequate prevention to exist. The first two subsystems center around intelligence gathering and dissemination, while the latter two sub-systems focus on identifying possible targets, protecting targets, and planning responses should a target be attacked. These processes are hierarchically interdependent in that failure in one area negates success in other areas. For example, failure to adequately assess targets means that response-protection systems likely will fail or be inadequate.

The prevention of terrorist attacks is a complicated matter. First, attacks can involve domestic citizens or groups of persons who reside within a country or those based in another country. For example, foreign as well as domestic right-wing terrorists have attacked targets in the United States. This complicates the identification and investigation of groups and incidents. Making matters worse, covert governmental operations are know to use terrorist-type attacks against civilian populations to advance political agendas. Second, attacks are secretive, making discovery difficult or impossible. In some cases, it is difficult to determine who or what group was responsible for an attack or if it was carried out by a government. Secrecy also complicates preventive efforts. Third, there is a host of potential targets in any country. Targets range from population centers or gatherings to institutions such as government buildings, financial institutions, communications centers, and transportation arteries. Finally, attacks can come in many forms. WMDs are a primary concern, but conventional weapons, bombs or explosives, are most often used and result in substantial destruction. Thus, prevention presents multiple challenges.

Threat Identification

In terms of threat identification, prior to the Iraq invasion, the greatest and most significant threat to the United States was al Qaeda, which was involved in the 9/11 attacks on New York and Washington. The invasion of Iraq has resulted in the United States being a focal point for a great deal of anger on the part of numerous terrorist groups. For example, Sheik Ahmad Bahr, the leader of Hamas, speaking at a mosque in Sudan recently called for all Americans to be killed (Staff, 2007). *Threat identification* requires that these groups be identified and monitored to identify specific and credible threats. However, it is a difficult task to identify all possible groups or cells with the intent to commit a terrorist act. Nonetheless, our intelligence and law enforcement agencies must endeavor to do so. While al Qaeda remains the foremost adversary in the public's mind, it is important that it has inspired others through its rhetoric and deeds. These groups do not have the ability as compared to al Qaeda and other older, more well-established terrorist groups, but they might be seen as opportunistic terrorists whereby they, a potential target, and the means intersect. These groups are dangerous from the perspective that they are isolated and generally intent on committing a single significant terrorist act. This *modus operandi* makes it difficult for authorities to discern their existence and strategies.

Threat Assessment

While identifying possible attackers, policymakers also must consider the types of potential attacks. Generally, WMDs have driven a substantial number of policy decisions. Policies have been promulgated that address all types of WMDs, but clearly some are more threatening than others (Union of Concerned Scientists, 2006). The possibility of terrorists using some WMDs is greater as compared to others. As an example, even though some bacteria are deadly, many cannot be passed on among citizens like viruses. Thus, viruses are more likely to be used as a terrorist weapon. On the other hand, certain viruses like small pox, while difficult to obtain, pose enormous dangers. There is limited access to viruses as a result of enhanced guardianship, and if obtained, the virus must be transported long distances in areas observed by a number of governments. This is not to say that these WMDs will not be used, but there are other weapons that can produce similar and are more easily obtained. There is a calculus regarding the potential use of various weapons that should be included in prevention policy.

Another factor to consider when gauging *threat assessment* is the ability of a group to acquire specific WMDs. Although many states possess a variety of WMDs, they are not inclined to provide them to terrorist organizations for fear that the weapons would be traced to the originating country, and the victim country would retaliate (Ackerman & Synder, 2002). Moreover, if a state were associated with the use of WMDs, it would result in numerous other diplomatic and strategic problems. To some extent, states have a vested interest in keeping WMDs out of the hands of terrorists. This is not to say that states will not supply these weapons, but the use of conventional weapons is far more probable.

Target Assessment

In terms of *target assessment,* the first step is to identify potential targets and, realistically, there are literally thousands of targets in any country. The two general categories of targets are critical infrastructure and population groupings. *Critical infrastructure* includes transportation, telecommunications, financial institutions, governmental buildings, medical care facilities, and petro-chemical facilities. *Population groupings* include permanent and temporary large numbers of people. Permanent populations include high-density populations located in cities and shopping areas where people congregate on a constant basis. Temporary population groupings include sporting and other cultural and entertainment events that draw large numbers of citizens. These population groupings are also vulnerable to attack.

The DHS is working to establish a national critical infrastructure database to guide prevention efforts. Moteff (2006) notes that the DHS established preliminary target guidelines that included criteria such as refineries with a capacity of 225,000 barrels of oil per day, commercial centers with a potential loss of $10 billion, or events or activities that involve more than 35,000 people. These types of targets should receive the highest priority. Even though the criteria are quite high and restrictive, they include hundreds of potential targets. Moreover, such criteria omit numerous targets that may be appealing to terrorists or which might result in substantial political consternation. For example, the New York World Trade Towers attack resulted in close to 3,000 deaths. The point is, target assessment has been quite problematic, and prevention must begin with the identification of possible targets, which is of great importance at the local level. It is the cornerstone for strategic planning. Response-protection decisions cannot be made without first identifying priorities.

In addition to attempting to build a database, the DHS has developed a formula for prioritizing potential targets. The formula [R=f(C,V,T)] where risk or R is equal to the function of consequences of attack C, the vulnerability of the target V, and the threat or risk of a terrorist attack T. The formula assumes that adequate information exists for each of the three variables. Consequences and vulnerability can be calculated, but oftentimes, threat is an unknown because planners do not have information about terrorists' intentions. Thus, security planners often must use consequences and vulnerability when prioritizing critical infrastructure protection.

Response-Protection Decisions

Once potential attackers, their capabilities, and possible targets and vulnerabilities are identified, decisions must be made concerning infrastructure protection. The national, state, and local governments must identify likely targets and develop *protection and response plans*—a process similar to the construction of the national critical infrastructure database. For the most part, these actions occur at the local level with state and federal support. On its face, this process appears to be rational and straightforward, but the process is so expansive and political, even for the smallest jurisdiction, it likely will not occur in a comprehensive, useable manner. As noted above, the DHS has been attempting to construct a national critical infrastructure database, but its efforts have fallen short of expectations. It is a rational process in an irrational world.

A factor complicating the protection of critical infrastructure is that the vast majority of these assets are privately owned. Although the government regulates many critical infrastructures, it only has tangential authority to mandate protective measures or changes. The various affected industries are not likely to volunteer changes, especially when they are costly and adversely affect corporation profits. Industry likely will lobby Congress to pay for any changes that contribute to infrastructure protection. Indeed, Gilmore (2003) notes that the presence of federal funding is a predictor of prevention status. In essence, government may not be able to enact the requisite changes for enhanced infrastructure protection and therefore reduce vulnerability for specific targets.

Homeland Security and Local Policing

National security has major implications for state and local governments and their police. Should a terrorist attack or major disaster occur, state and local police are among the first responders. They must control the situation and mitigate the damage as much as possible. Moreover, the federal government has approached homeland security from a top-down perspective. The federal government, to a large extent, has moved in the areas of airline security and there are mounting efforts in border security, although border security has little to do with reducing the threat to potential targets. For the most part, the federal government has concentrated on identifying and arresting terrorists or threat identification. There is substantial consternation over the role of local police agencies in protecting assets. However, we must "think globally and act locally." Essentially, attacks on a local asset can come from almost anywhere in the world, but local authorities will assume a large amount of the responsibility to prevent an attack and mitigate it if an attack does in fact occur.

Local agencies must develop strategies and policies to deal with security. As noted, it is the local agencies that will, at least initially, be responsible for protection and response. Realistically, states and local units of government have not received workable guidance relative to constructing and implementing such strategies. Currently, it appears that the states and local units of government are engaged in four security activities: (1) threat assessment via intelligence gathering, (2) critical infrastructure identification, (3) partnerships between law enforcement and critical infrastructure security personnel, and (4) public education.

Intelligence-Led Policing and Threat Assessment

Threat assessment at the local level implies that local police departments must become involved in identifying potential terrorists and suspect activities, although there is some debate as to whether terror intelligence gathering should be a federal function or involve state and local agencies (Thacher, 2005). The *Office of Community Oriented Policing* in the U.S. Department of Justice has developed and delivered a training program throughout the United States that focuses on intelligence-led policing (Carter, n.d.). The training program is designed to assist local law enforcement in enhancing criminal intelligence programs and expanding them to include information about potential terrorists. Most major police departments have some form of intelligence gathering capabilities or a crime analysis unit. Intelligence analysis differs from crime analysis, which often focuses on investigations or the deployment of police personnel or resources. "Intelligence is a formal process of taking information and turning it into knowledge while ensuring that the information is collected, stored, and disseminated appropriately" (Peterson, 2005:11). *Intelligence-led policing* requires the following steps:

1. Planning—requires an agency to identify the outcomes it wants to achieve from its collection efforts.
2. Collecting—intelligence analysis requires collecting and processing large amounts of information.
3. Analysis—converts information into intelligence. Analysis is quite simply a process of deriving meaning from data.
4. Dissemination—requires getting intelligence to those who have the need and the right to use it in whatever form is deemed most appropriate.
5. Reevaluation—is the task of examining intelligence products to determine their effectiveness (Peterson, 2005:7).

Of course, one of the major difficulties in collecting terrorist intelligence at the local level is that, unlike other organized crime groups, little is known about who might be a terrorist and their potential activities. For the most part, there is an absence of baseline data or information to guide intelligence and investigative activities. It is likely that both the police and public will evidence bias in the assumptions they make about terrorists and their activities, defaulting to popular media constructions that focus almost exclusively on international groups from the Middle East. Rhetorically, it is too late to gather information about possible terrorists once they have been identified as terrorists. Nonetheless, departments are encouraged to begin gathering information on "persons of interest." Source of intelligence information is generally collected from:

1. Physical surveillance (either in person or by videotape).
2. Electronic surveillance (trap and trace or wiretap).
3. Confidential informants.
4. Undercover operators.

5. Newspaper reports (now also Internet sources).
6. Public records (e.g., deeds, property tax records) (Peterson, 2005:6).

Intelligence-led policing is compatible and complementary with community policing. Community policing, to be effectively implemented, requires that police departments forge working relationships with citizens and communities (Kappeler & Gaines, 2008). This allows the police to better identify problems and react to them. Such relationships are conducive to the police collecting information about potential terrorist and suspect activities. In other words, the public becomes the "eyes and ears" for the police departments assisting them in gathering information for intelligence purposes.

Fusion Centers

In some of the major metropolitan areas, the FBI has partnered with state and local officials to form terrorism early warning groups or *fusion centers* (Sullivan, 2006). The fusion center provides overarching coordination of all response and counter-terrorism elements within a community. Essentially, information is fed into the fusion center from all agencies within a geographical area. The information is then analyzed. Once analyzed, threats or activity intelligence is generated and supplied to affected constituents. The teams include medical personnel and fire department personnel, as well as law enforcement personnel. The medical personnel can provide the fusion center with information about suspicious diseases or illnesses—an early warning system for a biological attack—and the firefighter personnel, can provide information about suspicious fires or chemical problems. This team model allows all possible information to be analyzed in a central facility. The fusion center also allows for more comprehensive planning and a better coordinated response should an event occur.

Critical Infrastructure Identification

At some point, local police departments will take inventory of the *critical infrastructure* within their jurisdictions. Communities will need to identify assets and develop response plans, especially given the numerous potential targets that may exist and because attacks on different types of targets present dissimilar challenges to the police and other first responders. Thus, it is important for local police departments to create their own a catalog or database of all possible targets in the jurisdiction, especially those that would result in significant damage if attacked.

The identification of these critical infrastructure assets serves two primary purposes. First, the database allows the department to develop response plans comprehensively. That is, a police department should have a response plan in place for all of these locations. Second, it results in focusing attention on areas that are of interest to potential attackers. Once assets are identified, the police department should focus intelligence operations near and around these locations. It is likely that if attackers plan to target an asset, he or she will conduct reconnaissance. Due vigilance may result in the suspects being identified before the act.

Police departments currently have critical incident response plans for natural disasters, airplane crashes, and major crimes such as hostage situations, etc. These plans include information about command and control, tactical responses, and use of other support agencies such as disaster, medical, fire, and chemical and radiological personnel. These plans are flexible because they can be used to deploy resources for a host of problems anywhere within a jurisdiction. Police agencies use these plans sparingly as there are only a few instances that call for their application. Donahue and Tuohy (2006) note that these plans often fail for a variety of reasons including uncoordinated leadership among the various responding agencies; failed communications including inoperability of communications systems and a lack of desire for agencies to communicate with each other; weak planning whereby plans are developed in a vacuum without the benefit of "real life" experiences; and resource constraints since most emergencies of any magnitude quickly strip a jurisdiction's resources to maintain a maximum response. Donahue and Tuohy note that some of these problems can be overcome through tabletop exercises or drills. However, generally when these drills occur, critiques are conducted piecemeal with each agency examining its response as opposed to the total response, and the critiques generally focus on what went right as opposed to identifying and documenting failures and problems. Police agencies must examine their critical incident response plans and ensure that they are comprehensive and applicable to threats. Pelfrey (2005) advises that planning is the most critical aspect of prevention. Many departments are well equipped, but do not have adequate plans in place.

Enforcement and Critical Infrastructure Security Personnel

Private security is mentioned here because it is a significant force in safety. Each year more is spent on private security than public police, and the private sector employs larger numbers of personnel than do public police departments (Morabito & Greenberg, 2005). Policing and private security are not necessarily mutually exclusive domains. For example, Green (1981:25) has defined the role of private security as, "those individuals, organizations, and services other than public law enforcement and regulatory agencies that are engaged primarily in the prevention and investigation of crime, loss, or harm to specific individuals, organizations, or facilities." This definition shows that there is substantial overlap between private security and the police. Private security continues to grow, and its role in public safety is enhanced as the result of the homeland security orientation.

Many private security personnel are assigned to guard much of the critical infrastructure in this country. Historically, there has been little cooperation or communication between the police and private security personnel even though they, to some extent, have parallel responsibilities. However, the need to secure critical infrastructure has changed this perspective. Now, police departments are encouraged to develop formal working relationships with private security firms. Such relations would: (1) improve joint response to critical incidents, (2) coordinate infrastructure protection, (3) improve communications and data interoperability, (4) bolster information and intelligence sharing, (5) prevent and investigate high-tech crime, and (6) devise responses to workplace violence (Ohlhausen Research Inc., 2004).

It is logical for the police to develop and formalize these relationships. The police and private security personnel have common goals and, to a great extent, the private security personnel are more informed about problems, critical points, and vulnerabilities. Many of these facilities have controlled access and activities, and therefore, security personnel are more likely to observe people and actions that are out of the ordinary or suspicious, and independently or in cooperation with the police, investigate them. At a minimum, the police should be aware of those facilities that are target hardened through private security and have procedures that include private security personnel when responding to those installations.

Partnerships with the police are critical. Some of the activities that must occur are:

1. cooperative training on the development and implementation of potential suspect profiles;
2. mapping potential targets in a jurisdiction to include security assets;
3. development and coordination of critical incident plans outlining responses to attacks and disasters;
4. better communication between law enforcement and the private security industry.

Although everyone in law enforcement is well aware of domestic terrorism, few departments have initiated comprehensive planning and programming. Threats of terrorism likely will remain for the next several decades, and the police must enhance their capabilities to respond to them. Private security will play a key role in this new priority.

Public Education

On May 9, 2007, the FBI arrested six suspects who had planned to attack and kill soldiers at Fort Dix in New Jersey. An alert clerk who was asked to copy a videocassette on to a DVD supplied the initial investigative lead. The cassette contained footage of the suspects training for their attack. The alert store clerk notified the police, which resulted in an extended FBI investigation and subsequent arrests. This incident exemplifies how law enforcement can obtain valuable intelligence information from the public.

There have been efforts to develop *public education programs* to achieve results similar to those in New Jersey. These programs encourage citizens to observe for and report suspicious persons and activities. These programs are similar to some of the drug and crime citizen reporting systems used in the United States. Here, police media campaigns urge citizens to call the police and report criminal or drug activities. Lyon (2002) advises that the police should develop programs in communities in close proximity to possible problems. One of the problems with these types of programs is that in some cases police departments receive many more citizen calls or tips than can be processed. When the police fail to respond to the calls, it results in lowered citizen evaluations of police performance and a reluctance to call the police in the future. Additionally, only a relatively few number of calls result in an investigation. The reporting of suspicious persons or possible terrorists will be even more problematic as citizens most likely will report racially or different persons who are not involved in any suspicious activities.

New York City and Homeland Security

The 9/11 attacks that occurred in New York City raised its awareness. City and police officials are acutely aware that their city remains a potential target for terrorists, and there are numerous potential targets including Wall Street, city, state, and federal buildings, sporting events, television and Broadway audiences, a mass transit system that carries more than 6.5 million passengers daily, and petro-chemical facilities. Consequently, New York City has made a number of changes to prevent future attacks, and to better enable the City to respond, should an attack occur. It is illustrative to examine some of the actions that New York City has taken, especially considering that the City likely has instituted more safeguards as compared to other cities.

First and perhaps foremost, the NYPD has reorganized to include several security elements within the department. The department has more than 37,000 police officers with approximately 1,000 assigned to "terrorist" duties. One of the tactics used by these officers is the "surge." Essentially, each day about 200 officers are sent to a specific location, usually a potential target. They surge in the area as a show of force. Officers observe for and investigate suspicious persons and activities. Along these same lines, the department has increased the number of bomb-sniffing canines and routinely deploys them throughout the City. These specially trained canines serve as a deterrent, as well as possibly locating explosive materials.

The City is proactively using counterterrorism tactics in the mass transit system. Each week, NYPD officers conduct more than 300 explosive screening deployments per week. Here, officers physically check bags, briefcases, and other containers or conduct an external swab of the containers for explosives reside. NYPD transit bureau supervisors are provided with radiation sensors and random radiological screening occurs on facilities. Various mass transit facilities are inspected daily to ensure that all alarms and access control systems are operational. Canine units are often used in mass transit to detect explosives (Falkenrath, 2007). Essentially, the department has substantially increased its efforts to deter attacks and to detect potential attackers.

The NYPD is actively involved in gathering intelligence about terrorist operations. The department created a Counterterrorism Bureau that has analysts and detectives examine terrorist organizations, potential terrorists, and bomb-making technology. The department has dispatched officers to a number of foreign countries to work with counterterrorism personnel. For example, when the Madrid bombing occurred, the NYPD officer in Israel was immediately dispatched to Madrid, collected intelligence information, and forwarded it back to the *New York Counterterrorism Bureau*. The *modus operandi* information collected overseas may be helpful in detecting and preventing attacks in New York City. The NYPD Coun-terterrorism Bureau cooperates with federal agencies, but is not dependent on them.

The City is also deploying an array of security hardware. The City is initiating the *Lower Manhattan Security Initiative*, which resembles London's Ring of Steel. Essentially, the City is installing security cameras throughout portions of the City. Eventually, there will be about 3,000 cameras installed with approximately 2,000 owned by private businesses. The London Metropolitan Police Department used similar cameras to identify suspects after the subway bombings in 2005. The City is also considering movable roadblocks that can be activated remotely, should a problem occur. In the future, the NYPD may install facial recognition programming to enhance the identification of suspected terrorists or criminals. The City is also installing radiation detection devices around the City's port of entry to screen cargo for radioactive materials.

New York is ahead of other major cities in the United States. The City has essentially gotten in front of the Department of Homeland Security. City officials see New York City as a plausible target and are attempting to prevent attacks. This move has resulted in a different NYPD. Counterterrorism is now one of its primary objectives. It is likely that other major cities will follow New York's lead. For example, Pelfrey (2007) surveyed departments in South Carolina and found that many departments, of all sizes, had begun implementing homeland security programming.

Summary

A primary role of the police is the protection of life and property. As economic conditions change, the demands on the police institution have also changed. The geographic areas, individuals, and behaviors policed during each economic revolution have revolved around threats to the expansion of capital. The fact that a new economic revolution is underway is leading to a shift in American policing. Its geographic focus, the individuals, and the behaviors policed are changing. For example, police profiling of the 1990s largely involved drug-law violators selectively targeting minority males (Cole, 1999); now, that bias has shifted to young men of Middle Eastern descent. The policing of terrorism and the rhetoric of this agenda are leading to a federalization of municipal police agencies and an internationalization of federal agencies. Local and state law enforcement organizations are increasingly funded by federal sources while simultaneously experiencing a reduction in local and state monies. They are increasingly trained and provided equipment designed to fight an anti-terrorism cause. Historically, international crime was of little interest to American policing, but this is changing. Globalization is the driving force in the new international interest of the American police institution.

The newly identified targets by police, and the expanded geographic focus, are based on the identification of the groups and regions thought to be most threatening to the United States. As seen in New York City, this new problem results in a shift in organizational priorities and re-organization. It means that these new responsibilities must be integrated into departments. It is problematic in that for the most part, police departments are expected to pursue traditional goals and objectives. In the end, police departments are becoming more complex as the number of public and government expectations increase.

References

Ackerman, G. and K. Moran, (n.d.) "Bioterrorism and Threat Assessment." Stockholm, Sweden: Weapons of Mass Destruction Commission. http://www.wmdcommission.org/files/No22.pdf

Ackerman, G. and L. Snyder (2002). "Would They if They Could?" *Bulletin of the Atomic Scientists*, 58(3):40–47.

Anti-Defamation League (2007). *The Ku Klux Klan Rebounds*. Washington, DC: Author's Web site.

Appadurai, A. (1990). "Disjuncture and Difference in the Global Cultural Economy." *Public Culture*, 2(2):1–24.

Carter, D. (n.d.) *Law Enforcement Intelligence: A Guide for State, Local, and Tribal Law Enforcement Agencies*. Washington, DC: Office of Community Oriented Policing Services.

Cole, D. (1999). *No Equal Justice: Race and Class in the American Criminal Justice System*. The New Press: New York.

Department of Homeland Security (2006). *National Infrastructure Protection Plan*. Washington, DC: Author.

Donahue, A. and R. Tuohy (2006). "Lessons We Don't Learn: A Study of the Lessons of Disasters, Why We Repeat Them, and How We Can Learn From Them." *Homeland Security Affairs*, 2(2):1–28.

Falkenrath, R. (2007). Prepared statement of testimony before the Committee on Homeland Security, United States House of Representatives, March 6, 2007.

Freeh, L. (2001). Statement for the Record on the Threat of Terrorism to the United States before the United States Senate Committees on Appropriations, Armed Services, and Select Committee on Intelligence, May 10, 2001.

Gilmore, J. (2003). "Forging America." *New Normalcy: Security our Homeland, preparing our liberty*. Arlington, VA: Rand.

Green, G. (1981). *Introduction to Security*. Stoneham, MA: Butterworth.

Henry, V. and D. King (2004). "Improving Emergency Preparedness and Public-Safety Responses to Terrorism and Weapons of Mass Destruction." *Brief Treatment and Crisis Intervention*, 4(1):11–35.

Hobijn, B. (2002). "What Will Homeland Security Cost?" *Economic Policy Review*, 8(2):21–33.

Hoffman, B. (2004). "Consumerism." In Bruce Mazlish and Akiria Iriye's (eds.) *The Global History Reader*. New York: Routledge.

Jenkins, B. (1984). "The Who, What, When, Where, How, and Why of Terrorism." Paper presented to the Detroit Police Department Conference on Urban Terrorism: Planning or Chaos? November.

Kappeler, V.E. and L. Gaines (2008). *Community Policing: A Contemporary Perspective*, Fifth Edition. Newark,

NJ: LexisNexis Matthew Bender.

Kleiman, M. (2004). *Illicit Drugs and the Terrorist Threat : Causal Links and Implications for Domestic Drug Control Policy*. CRS Report to Congress. Washington, DC: Congressional Research Service.

Laqueur, W. (1987). *The Age of Terrorism*. Boston: Little, Brown.

Lash, S. and J. Urry (1994). *Economies of Signs and Space*. Newbury Park, CA: Sage.

Lash, S. and J. Urry (1987). *The End of Organized Capitalism*. Madison, WI: University of Wisconsin Press.

Lyon, W. (2002). "Partnerships, Information, and Public Safety." *Policing*, 25:530–543.

Mahoney, J. (2007). "Activists Confront Controversial Educator: Demonstrators Charged as Scuffle Erupts Over Ex-teacher Tied to White Supremacists." *The Globe and Mail*, April 20, 2007.

McCaffrey, B. and J. Basso (2003). "Narcotics, Terrorism, and International Crime: The Convergence Phenomenon." In R. Howard and R. Sawyer's (eds.) *Terrorism and Counterterrorism: Understanding the New Security Environment*. Guilford, CN: Dushkin, pp. 206–221.

McCloud, K.A. and A. Dolnik (2002). "Debunk the Myth of al Qaeda: Its Size and Reach Have Been Blown Out of Proportion." *Christian Science Monitor*, May 23, 2002.

Morabito, A. and S. Greenberg (2005). *Engaging the Private Sector to Promote Homeland Security: Law Enforcement-Private Security Partnerships*. Washington, DC: Bureau of Justice Assistance.

Moteff, J. (2006). *Critical infrastructure: The National Asset Database*. Congressional Research Service Report. Congressional Research Service.

National Intelligence Council (2007). *National Intelligence Estimate*. "The Terrorist Threat to the U.S. Homeland." Washington, DC: Office of the Director of National Intelligence.

O'Hanlon, M. (2003). "Protecting the American Homeland: One Year on." Washington, DC: Brookings Institution.

Ohlhausen Research Inc. (2004). *Private Security/Public Policing: Vital Issues and Policy Recommendations*. Alexandria, VA: International Association of Chiefs of Police.

Pelfrey, W.V. Jr. (2007). "Local Law Enforcement Terrorism Prevention Efforts: A State Level Case Study." *Journal of Criminal Justice*, 35:313–321.

Pelfrey, W.V. (2005). "The Cycle of Preparedness: Establishing a Framework to Prepare for Terrorist Threats." *Journal of Homeland Security and Emergency Management*, 2(1):1–21.

Perrow, C. (2006). "The Disaster After 9/11: The Department of Homeland Security/and the Intelligence Reorganization." *Homeland Security Affairs*, 2(1):1–32.

Peterson, M. (2006). *Intelligence-Led Policing: The New Intelligence Architecture*. Washington, DC: U.S. Department of Justice, Office of Justice Programs.

Staff (May 1, 2007). "Top Hamas Official: 'Kill All Americans.'" *Jerusalem Post*. http://www.jpost.com/ servlet/Satellite?pagename=JPost%2FJPArticle%2FShowFull&cid=1178020746583

Stables, E. (2007). "Squabbles between ICE, FBI Affecting Terrorism Investigations." *CQ TODAY*, Aug. 15, 2007.

Sullivan, J. (2006). "Terrorism Early Warning Groups: Regional Intelligence to Combat Terrorism." In R. Howard, J. Forest, and J. Moore's (eds.). *Homeland Security and Terrorism*. New York: McGraw-Hill, pp. 235–245.

Thacher, D. (2005). "The Local Role in Homeland Security." *Law & Society Review*, 39:635–676.

Union of Concerned Scientists (2006). "Nuclear Terrorism." http://www.ucsusa.org/global_security/nuclear_terrorism/

United Nations Panel (2007). UN Reform. New York City: UN Web Services Section, Department of Public Information, March 21, 2005.

Wermuth, M.A. (2005). The Department of Homeland Security: The Road Ahead. Testimony before the Senate Committee on Homeland Security and Governmental Affairs, January 26, 2005.

White, J. (2006). *Terrorism and Homeland Security*. Belmont, CA: Wadsworth.

Wise, C. (2006). "Organizing for Homeland Security After Katrina: Is Adoptive Management What's Missing?" *Public Administration Review*, 66:302–318.

Woodbury, G. (2005). "Measuring Prevention." *Homeland Security Affairs*, 1(1):1–9.

Chapter 4

"Fragmented" Security: Protecting the Homeland in the Twenty-First-Century United States

By Jeff Maskovsky and Hilary Cunningham

> Emergency preparedness is no longer the sole concern of earthquake prone Californians and those who live in the part of the country known as "Tornado Alley." For Americans, preparedness must now account for manmade disasters as well as natural ones. Knowing what to do during an emergency is an important part of being prepared and may make all the difference when seconds count.
>
> —Department of Homeland Security, Ready Program, 2007

In the aftermath of 9/11, the Bush administration made clear its objective to organize a security matrix around federal, state, local, and private organizations "whose primary functions"—as the White House's 2002 National Strategy for Homeland Security noted—traditionally had been "unrelated to national security." The planned and deliberate linkage of "unrelated" organizations across levels of government and across the public-private divide to the national security apparatus of the U.S. government signaled that homeland security was meant to be a new organizing rubric for political authority and territorial control in the United States.

In good liberal democratic fashion, public deliberation during the Bush years—on the PATRIOT Act in particular—was largely restricted to calls for new and better security in the United States, on the one hand, and worries over the implications of the Bush administration's new counterterrorism measures for civil and human rights, on the other. Like many progressive academics and people of conscience, we shared—and still share—the concern that the latter would be sacrificed in the name of the former. Yet then, and now, we are not satisfied with the limited range of positions staked out in the mainstream debate over the proper balancing of security with rights in the new age of global terrorism. To move us past the security versus liberty framework, we will focus in this chapter on the implications of homeland security as a broad, ideological rubric reorganizing the United States into a landscape of potentiality interconnected threats and emergencies that require new kinds of surveillance, security, and protection. We will explore the contours of this new security landscape by pursuing connections between "the city" and "the border"—two prominent sites that, like ports and airports, have experienced the imposition of new surveillance and security measures since 9/11.

We direct our attention to cities and international borders in particular because they are notoriously difficult to govern. Cities have long been considered a threat to U.S. national identity and purpose: havens for immigrants, unions, African Americans, gays, and unmarried women, they are too dark, too foreign, too criminal, and too deviant to be considered safe and secure for the U.S. mainstream. As "ulcers on the body politic" (Jefferson) and incubators of "moral contagion" (Riis), cities require by their very nature special governmental measures for the maintenance of the social order. International boundaries, for their part, are often regarded as a litmus test for determining how strong a nation truly is. Improperly guarded, borders can signal not only an impotent state but also an entire nation at risk. The intrusion of foreign elements (having penetrated the nation at its first line of defense) thus endangers the entire body politic, and the fact that many of these elements seek out and find haven in urban centers has provided a long-standing link between the security of the nation's cities to that of its perimeters.

Prior to 9/11, both cities and borders were considered to be undergoing considerable changes in the context of neoliberal globalization. The global flow of finance, technology, and labor was said to "denationalize" cities, while at the same time making national borders anachronistic to the new global order. After 9/11, perspectives on the so-called global nature of cities and the growing irrelevance of national borders changed considerably as homeland security recast them as key sites within a new geography of security and endangerment. In this chapter, we are not going to address the extent to which new coercive measures are effective in combating global terrorism. We are interested, instead, in assessing their implications for national security and civil liberty and also for the equally important and long-standing issues of citizenship and social inequality. We shall argue that homeland securitization—as a distinctive outgrowth of national security—is accelerating and amplifying some of the most undemocratic, racist, and class-polarizing tendencies that have come about in recent decades as the United States has moved away from a political economy based on Keynesian managerialism toward one based on neoliberal entrepreneurialism.[1] Indeed, it is precisely the way in which homeland security is now being grafted onto—and enabling the expansion of—neoliberal surveillance and policing strategies that makes it so dangerous and that therefore requires our scholarly and political attention.[2]

The United States in Crisis and the Emergence of the Homeland

Following the terrorist attacks of 2001, the Bush administration embarked on a massive reorganization of the government bureaucracy through the creation of the Department of Homeland Security (DHS). Twenty-two former government agencies were absorbed into the DHS, giving the new agency an official staff of over 180,000. Added to this roster were thousands of U.S. community groups or citizen groups consisting of patriots who were to act as the eyes and ears of the nation. Arguing that securing the homeland from terrorist attacks was an "exceedingly complex mission" requiring "a focused effort from our *entire society*" (emphasis added), the DHS sponsored the federal Citizen Corps coordinating around four programs: Neighborhood Watch, Volunteers in Police Service (VIPS), Community Emergency Response Team (CERT), and Medical Reserve Corps.[3] The objective of these projects was to establish a nation that would be aware, alert, and ready to respond and to nurture a citizenry that would be better prepared "for preventing and handling threats of terrorism, crime and disasters of all kinds." According to the Citizen Corps program, the benefits to the communities that embrace responsibility and emergency training would be "a greater sense of security, responsibility, and personal control of their community and the hazards they may face," as well as building "community pride and patriotism while preparing us all for helping others in a crisis."[4]

Although the DHS clearly sought an expansion of preexisting relationships between municipal, state, and federal political jurisdictions and underscored the role of the individual citizen in a new parapolicing capacity, the term *homeland security* was not an entirely unfamiliar one on Capitol Hill prior to September 2001. According to some researchers, the term surfaced in connection with security policy as early as 1995 and had become widely used in congressional documents by 1998 (Beresford 2004, 1). Nor were civilian defense initiatives something new to the United States. The civilian-based rescue, first aid, and firefighter volunteerism in place in the country since at least the early 1940s provided the basis for the new Citizen Corps, as well as projects such as the Terrorism Information and Prevention System (TIPS)—the latter not unlike Cold War programs asking U.S. citizens to support the Federal Bureau of Investigation (FBI) fight against communism by handing over any relevant information on "espionage" and "subversive activities" (Beresford 2004, 2).

Clearly, many organizational and cultural precedents for homeland security were in place long before 9/11, but the discursive shift from "national defense" to "homeland security" was not without ideological significance. The term *homeland* is a complicated one, as it has been used in both fascist and liberation movements, and its inherent linkage between a specific population and a geographic place has historical connections to both totalitarian and emancipatory politics. In the late 1990s, the the word was used in U.S. policy circles to distinguish threats directed at U.S. civilians living in the contiguous United States (i.e., "at home") from those aimed at military personnel (as well as their families and staff) stationed "abroad." America at-home, under this usage, could no longer be regarded as a place where U.S. citizens lived at a "safe distance" from U.S. geopolitics—a sense that was only confirmed by the events of 9/11 and reiterated in frequent comparisons of the World Trade Tower attacks to the Japanese attack on Pearl Harbor on December 7, 1941. The possibility of blowback, then, was on the U.S. government's national security radar screen long before 9/11 (Johnson 2004).

In becoming a homeland, the United States was also "reethnicized" in ways that built upon deeply racist and nativist attitudes but nevertheless underscored the nation as a population with a distinctive identity, history, values, and way of life *in contrast to others.* One could argue, in fact, that the U.S PATRIOT Act was the principal means through which the homeland was "ethnicized" via the category of "patriot," which has historically coded as both white and Christian. Moreover, although this shift to homeland did not mean that the United States gave up any of the rights and entitlements of being a sovereign state within a system of nation-states, it did pry open the door to questions about special protections, emergency measures, and unprecedented forms of security—in short, to "homeland jurisdictions" that could not be properly addressed in terms of traditional Westphalian state rights and international legal regimes.

Hence, just as 9/11 made Americans newly vulnerable at home, so, too, did the homeland create new categories of membership and exclusion in which political and civil rights could be separated (for example, in the designation of illegal noncombatants) and in which certain persons could be subject to unusual and enhanced forms of security and surveillance (as in extraordinary renditions and the use of torture as an interrogation technique). Real patriots, of course, were not to be subject to these new forms of security. Yet clearly, the U.S. homeland—with its "porous" border with Mexico to the south, its "undefended" northern border with a nation whose immigration and refugee laws were woefully lax, and its cities teeming with illegal migrants—was in a bad state of affairs. What was needed was not only better, more secure cities, military bases, native reserves, ports, borders, and so on but also newly trained, *real Americans* who could both read and help to police these vulnerable jurisdictions in a proper manner. It is no surprise, then, that 9/11 was subsequently renamed and memorialized as "Patriot's Day."

The startling alacrity, however, with which the United States became a homeland—both in terms of the sweeping post-9/11 legislation and the unprecedented reorganization of government bureaucracy—could not have occurred without the widespread consensus, orchestrated largely by politicians and the media, that the country was *already in a state of severe crisis* and that something drastic had to be done. The crisis in its cities and at its borders, in fact, had been a long-standing and cumulative one. The mid-twentieth-century restoration of the U.S. economy had been accomplished, in large measure, by a confluence of geopolitical and domestic circumstances that could not be sustained indefinitely. By the 1970s, rising oil prices had exacerbated a brewing crisis in corporate profitability that sent shock waves through the economy, plunging the country into a pronounced period of economic decline, shattering the New Deal consensus, and setting the stage for the new business-led consensus that globalized the economy under neoliberal precepts. Moral panics about U.S. cities and borders as well as immigration policies were an essential part of this transition, and long before the terrorist attacks on 9/11, they spurred a neoliberal security crisis.

Locking Down the City

In the 1980s and 1990s, U.S. cities tended to be represented (on both the left and the right) as repositories of a redundant labor force, with groups of poor people, particularly poor people of color, restricted to deteriorating inner cities (Massey and Denton 1993; Wilson 1987; see also Goode and Maskovsky 2001; Segrue 1998). In this formulation, the city appears as a victim of the suburbs, of economic change, and of federal neglect. Policy discussion and the popular imagination subsequently became obsessed with the depressing, overwhelming images of decay, civil disturbance, and fiscal bankruptcy that characterized the flight of manufacturing and population from many industrial cities from the 1950s to the 1980s. The focal point for these images was clearly the 1970s, punctuated by the aftermath of the urban riots in the late 1960s and New York City's fiscal crisis of 1975.

In reacting to—and helping create—these images, the right has been concerned to demonstrate how "moral bankruptcy" and the "culture of poverty" have produced an urban "underclass" isolated from market forces, supposedly unwilling to take the responsibility or action necessary to improve U.S. cities. The left has used the same negative urban imagery to develop structural arguments to combat the "up by the bootstraps" (and then later "family values") ideology deployed by the Right to oppose the modern U.S. welfare state.

But since the mid-1990s, there have been significant changes in the economies and geographies of many U.S. cities, and these changes have significantly altered public perceptions of the urban landscape and of poor people within it. The Keynesian orientation of the immediate postwar period was replaced in the mid-1970s with a neoliberal orientation, one that involved the state-coordinated return of banking and real estate investment to the U.S. urban core and of population in the form of gentrifying middle and upper classes. By the 1980s, urban planners and boosters across the United States had become hyperconscious of the dangerous and backward reputation of U.S. cities and devoted themselves to reversing it by offering urbane and cosmopolitan images of the city for potential tourists, investors, and conventioneers. Different cities required different marketing strategies, of course, but nearly all municipal elites have embraced image enhancement as an important part of their postindustrial growth strategies. This situation had particular implications for the government of putatively "ungovernable" urban populations.

It is in the admixture of economic development policy and image enhancement that we may begin to understand the obsession over security in cities and the violence and state coercion that has been imposed to achieve that security. Since well before 9/11, the strong arm of the neoliberal state has been enormously important in creating the conditions for the recent upsurge in capital investment described earlier. Beginning in the 1970s, drug users, welfare mothers, undocumented workers, and others, typically coded as a racialized urban underclass, were targeted through policies such as the war on drugs, sidewalk ordinances, workfare, and other draconian and revanchist measures designed to remove them from public space and force their compliance with new postwelfare work regimes. In the neoliberal city but not of it, they were the targets of new urban security measures that Mike Davis famously equated with U.S. military policies in Vietnam (Davis 1992).

Yet militarization is not the only strategy used against the poor. A decade-long inner-city real estate boom combined with welfare "reform" to remake the inner city, ideologically and materially. Since the mid-1990s, the urban "underclass," no longer a metonym for the city itself, has frequently been offered redemption in the new call for civic engagement that can be heard across the urban landscape. No longer simply vilified as deviant and dependent, the urban poor are now actively encouraged to join cities in their struggles to become more attractive and viable places for private investment. Indeed, despite its appalling record in alleviating urban poverty, neoliberal restructuring produced opportunities for urban belonging and citizenship for the inner-city poor.

The city of Philadelphia, for example, embraced the "Neighborhood First" approach, which seeks to devolve authority, responsibility, and accountability to poor neighborhoods as part of a citywide strategy of urban revitalization. Poor neighborhoods were described in glossy brochures and marketing materials as the "hometowns of the 21st century," and their residents were called "good people with good values" (Office of Housing and Community Development 1998). In a highly publicized marketing analysis commissioned and disseminated by the city planning commission, "diversity" was singled out as a strong selling point for what were once considered to be racially segregated, immiserated inner-city neighborhoods. Diversity can, the report tells us, help to offset the lack of desirability of neighborhoods burdened by a dearth of city services. In this formulation, the urban poor were vested with "real authority" in a coordinated, city-funded system of community-based planning and service delivery organizations. Accordingly, the resident-volunteers who ran these organizations were charged with a variety of responsibilities, from community policing to block cleanups. And although these efforts tended to have almost nothing to do with the backroom deals that influenced the pattern of private investment, they nonetheless placed enormous burdens on the poor and disenfranchised to manage the problems produced by chronic disinvestment in their communities in a period of welfare state decline.

Making the cities of the homeland more secure followed a similar pattern. In 2003, the federal government established the Urban Areas Securities Initiative, which in its first year pumped $100 million into large urban areas to help local governments prepare for and respond to terror threats.[5] Public funds were also directed to "security research," a burgeoning field so lucrative that it has allowed cities such as Cleveland and Buffalo to establish security think tanks.[6] The federal government helped to establish thirty-eight state and local "fusion centers" designed to blend local law enforcement activities with intelligence information gathering and other counterterrorist measures. Local police forces added antiterrorist units, and a new global "Secure Cities" movement promoted the safety of cities in the age of global terrorism. Another interesting development with implications far beyond urban areas is the synergy that emerged between big data-collection and data-management companies and the U.S. government, as surveillance collected to track consumption patterns was given to the government to monitor citizens and noncitizens alike—here, a brave new world meets 1984.

Crucial to these efforts was volunteerism by urban residents, which DHS policies sought to leverage in order to reduce the threat of terrorism. In Philadelphia, for example, the city government established a Web site where residents can report anonymously on suspicious activities, and in New York, a multilingual antiterrorist ad campaign encouraged subway riders to speak up if they witnessed something questionable: "If you see something, say something."[7] Moreover, security experts were dispatched across the United States to help urban residents convert their Neighborhood Watch programs, a mainstay of urban community policing since the 1980s, into homeland security programs. And of course, the federal government rounded up and incarcerated numerous "suspected terrorists," many of whom were held indefinitely in captivity in detention centers in urban areas such as Queens, New York, and Miami, Florida.

The point here is that the most overtly coercive forms of state violence—those designed to sequester the poor through military-style "containment" strategies (Davis 1992)—did not operate alone. They worked in tandem with forms of neoliberal self-governance that place enormous burdens and responsibilities on the poor segments of the urban population. The net effect of these twin strategies—of simultaneous containment and acculturation—has been the polarization of citizenship within racially and ethnically defined urban populations. Indeed, well before 9/11, the coherency of the body politic and the "American way" has been maintained in large measure through the invocation of an internal Other—not necessarily a raced "other" (though it is often so) but an "other" of the market. This internal villain, this "other" of the market, is to be either converted into a good citizen or destroyed. The ideology of private investment and economic growth seeks to accomplish the former. The law-and-order state, with its prison system, brutality, and invocations of safety and community hygiene, seeks to do the latter.

This ideological assault has widened considerably since 9/11, as an even more dangerous and disabling connection is now made between terrorists and what remains in cities of the urban poor. Indeed, a neo-McCarthyite thuggery can be seen in the surveillance and incarceration of Muslims, Arabs, Arab Americans, and others who are viewed as terrorist threats to the wounded nation. Placed in a state of exception that suspends the rights of suspects, the war on terror resonates strongly with the assault on rights and the degradation of citizenship waged in the name of the war on drugs. Liberal pundits have been somewhat perplexed by the profound indifference on the part of much of the U.S. public to the assault on the rights of Arabs and Muslims inside and outside the United States. The nation's long-standing imperial xenophobia goes far in explaining the widespread tolerance of—and occasional enthusiasm for—the Bush-era policies on torture, rendition, and secret prisons. At the level of popular culture, it also explains the popularity of the TV show *24*, which repeatedly presents to its protagonist—and to the viewing public—contrived scenarios that require an immediate choice between preserving the rights and dignity of a known terrorist or preventing the massive loss of innocent life.

But there is more to it. The long-waged assault on the rights of urban African Americans has in many ways paved the way for the new coercive policies associated with homeland securitization. Indeed, the sacrifice of freedom for safety has long been seen as a political and ideological requirement for the maintenance of the U.S. urban order. The treatment of protestors at the Republican National Conventions in Philadelphia in 2000 and in New York City in 2004 illustrates how efforts to curtail criminal activity in inner-city neighborhoods broadened in recent decades to stifle dissent and suppress protest movements in the name of urban image enhancement (see Maskovsky 2003). Moreover, it is possible to see the black inner city of the 1970s, 1980s, and 1990s as the laboratory where the sinister provisions of the war on terror were invented and refined. The widely noted fact that some of the torturers at Abu Ghraib were Pennsylvania prison guards is but one example of this uncomfortable convergence.

Locking Up: Securing the Border

But could U.S. cities take on the new challenges of homeland security through new policing and surveillance strategies as well as through insidious processes of reconstituting poverty as community responsibility in the context of a nation with porous borders? As international relations scholars and political geographers have long pointed out (Sibley 1995), the security of a nation-state is primarily framed within the context of territorial boundaries and their protection. Since the 1970s, there has been widespread conviction among cultural pundits, politicians, economists, and academics that national borders will disappear in the face of a globalizing world. Such rhetoric was commonplace in the United States both before and after the implementation of the North American Free Trade Agreement (NAFTA) and the commencement of what was hailed as a new integrated economy among Mexico, Canada, and the United States. Yet despite the claims to the emergence of a borderless world brought about by flexible production, electronic finance, and the liberalized movement of goods with free-trade zones, the 1990s saw not only the implementation of NAFTA but also the beginnings of a dramatic militarization of the U.S.-Mexico border.

Beginning in late 1993, the U.S. Immigration and Naturalization Service along with the Border Patrol began to implement a border enforcement strategy that concentrated state resources (including Border Patrol agents, reinforced fencing, and surveillance technology) at four key, "high-traffic" crossing points for illegal entry. Starting in Texas and spreading eventually to California and Arizona, the plan concentrated enforcement agents in larger transborder cities such as Ciudad Juarez—El Paso, Tijuana—San Diego, and Ambos Nogales. The idea behind the strategy was to make it more difficult for illegal entrants from Mexico to cross into the United States. As a result, the strategy was called "prevention through deterrence," based on the assumption that if urban crossings were made difficult, then the harsh geography of the region would "do the rest."

Although the new strategy did indeed make illegal border crossing more difficult, no convincing evidence surfaced to demonstrate that it actually deterred undocumented migration to the United States. Among the neonativisms that emerged in the 1990s, we can point to the changing situation confronting one anti-NAFTA constituency—white (male-headed) families who were experiencing a decline in their economic base to the point that, in their eyes, they had been downgraded to the standing of working Latinos and blacks, many of whom were new immigrants. All of these individuals, including whites, were increasingly becoming what came to be known and talked about in academic and policy circles as "the working poor." This development started in California and spread quickly to the South and Southwest, conflating race and class dynamics at a moment when they were actually becoming separated (at least when it came to official labor markets). This situation was, then, a backlash against the loss of a racial advantage in low-wage labor markets. Labor and free-trade politics at the

U.S.-Mexico border were further fueled by racist legislation in California in 1992 that stimulated anew the debate about the "immigration crisis" in the United States. Here, the "border" emerged once again as a central site of contested claims about citizenship and medical and educational entitlements, as well as the status of children born to illegal migrants while residing in the United States.

Predictably, the events of 9/11 exacerbated debates about migration and border enforcement, but they also established a set of connections between the nation's illegal immigration crisis (as distilled in an "insecure" border) and its mandate to respond to threats in the homeland. "Securing our borders" thus became the new mantra of the homeland, to such an extent that discussions and debates about migration policy could not be held outside of debates about "securing our borders."

Post-9/11 immigration legislation and policies significantly expanded the practices of border enforcement to include, for example, an emphasis on workplace surveillance, as well as punitive legislation to fine employers who did not conduct proper document checks of their employees. Both Arizona and California proposed legislation that would force emergency hospital personnel and school principals to check immigration documents, and California rekindled a series of long and acrimonious debates about preventing undocumented migrants from obtaining driver's licenses. In the post-9/11 climate, then, the so-called porous U.S.-Mexico border of the 1990s was reconceptualized and redeployed so that it encompassed not only the borderline itself but also sites typically associated with the U.S. interior and particularly U.S. cities. The Immigration and Customs Enforcement (ICE) house raids, especially those commencing in 2004, have also underscored the ability of the state to enforce the "border" in the country's urban spaces and across public-private divides.[8]

Fears about the trajectories of illegal border entry have also dominated the U.S. desire to enforce the northern border with Canada—an international line that for much of its history has been not only unguarded but also relatively insignificant in Canada and U.S. foreign policy deliberations (Andreas and Biersteker 2003). Post-9/11, this border became newly problematic in view of Canada's seemingly liberal immigration procedures as well as its nationalist ideology of a tolerant multiculturalism. As Canada and the United States struggled to put a "smart" border in place—that is, a border where "high risk people and goods" could be identified while "expediting the flow of low risk goods and people" (The Smart Border Declaration 2001)—U.S. politicians such as then-senator Hillary Clinton began to argue for the need to coordinate U.S.-Canada immigration and refugee policies and focus attention on what were deemed "high-risk areas" of the Canadian border. Risk in this sense was not just about unsupervised entry as a problem in itself; it was also about the vulnerability of New York State's urban populations to biological and other forms of terrorism as a result of Canada's "slack" border enforcement. In December 2002, for example, CNN reported that five Al Qaeda suspects had crossed into the United States via Canada and were quite possibly intent on disrupting New York City's large, outdoor New Year's Eve festivities. Although the report turned out to be false, there was considerable speculation that the terrorist suspects had crossed at the St. Regis Mohawk Tribal Reservation, a territory that spans the U.S.-Canada border and involves multilevel tribal, state, provincial, and federal policing bodies.

These pre- and post-9/11 developments in U.S. border enforcement and immigration policy, when looked at together, suggest that understandings of what a border is have been significantly reconstituted in the United States, to such an extent that U.S. *borderlines* have become more of a *borderland* or, in other words, a new kind of *border regime* (Euskirchen, Lebuhn, and Ray 2007), one that connects the nation's interior spaces to its edges through a network of interconnected risks. In the homeland, points of entry immediately connect to the circuits through which unwanted elements can move deep into the nation's interior regions—deep into its cities and the kinds of anonymity, invisibility, and access to informal livelihoods that cities can offer to illegal entrants.

The focus on risk in these new border regimes involves, in particular, refugees, who have been reframed in terms of fraud and criminality. As criminologist Anna Pratt observes, rather than concentrating on indentifying and responding to need, neoliberal immigration policies have instead turned to an obsession with those who cheat and abuse the system (Pratt 2005). In Canada, this approach has been expressed as a "closing of the back door" to system abusers in order to keep the front door open to "legitimate" refugees and immigrants. The criminalization, in particular, of asylum thus links border security to new kinds of neoliberal audits of the immigration, citizenship, and visa systems, all of which have been inextricably tied to a securing of the border.

As a result, it is not only the borderline that became subject to an aggressive securing but also the homeland's points of transit (airports, bus terminals, train stations, and subways), typically located in urban centers, as well as its approach to immigration itself. The "border," it seems, has come to reside throughout the U.S. homeland—at both its edges and in its cities—so that the subway car (where one might just see something) apparently requires the same kind of vigilance as does a port of entry.

Conclusion: "Fragmented" Security and the Privatization of Risk

In the aftermath of 9/11, we saw the emergence of a distinctive conjuncture between national security and neoliberalism. Notable for its adherence to—and enhancement of—the values of flexibility, minimalist government, private sector entrepreneurship, and citizen volunteerrism that are the hallmarks of the neoliberal agenda, the Bush administration's security regime not only reflected the longstanding inequalities on which U.S. neoliberalism has been based but also produced new inequalities as the country was reorganized to accommodate its new search for security. Although articulated largely through the vision of a homeland, with its attendant assumptions about the homogeneity and unity of a given (ethnic) population, the Bush administration's homeland security matrix was not designed to be a unified system that distributes risk and security equally. Instead, what was envisioned and produced was a fragmented system, designed both to distribute risk and security in unequal ways and to identify and isolate what are deemed particularly high-risk individuals, resulting in new consequences for different subject populations, as our analysis of Bush administration securitization efforts makes clear.

The dangers inherent in a fragmented security regime are by no means unprecedented (Cold War security was itself fragmented, though along different political and social lines). Yet many of its most serious dangers were ignored, even by some of the most outspoken critics of the Bush administration. Certainly, the introduction of coercive measures into new domains of U.S. life and the permanent abrogation of long-standing freedoms from government intervention and surveillance pose dangers for everyone living in the United States today. And this is precisely why civil libertarians decried these developments, worrying that they, along with warrantless wiretapping, rendition, waterboarding, and other Bush administration measures of dubious constitutional legality, undermined core American values in the name of the war on terror.

Of equal importance, there is no reason to believe that the sacrifice of liberty will actually make Americans more secure. The Bush administration established a new security regime that was rather narrowly focused on counterterrorist surveillance, with disproportionate legal and public policy attention and funding devoted to surveillance and policing strategies designed to identify terrorists and to block their entry into the country. Homeland securitization has thus far worked in many respects to direct attention away from the broad range of serious security issues facing the country today. (The Bush administration budget for 2008 proposed $12 billon for border security and immigration enforcement but only $2.2 billion for homeland security protection to *all* U.S. cities, and it actually sought cuts in other vital areas of homeland security funding that are not directly related to counterterrorist measures.) What about natural disasters, infectious diseases, food contamination, environmental toxins, global warming, and infrastructural decay, to name a few mounting security threats, that are making the United States much less secure? Clearly, without a sea-change in the way the United States understands threats and security, no one could be surprised by a repeat of what happened in New Orleans and on the Gulf Coast in the aftermath of Katrina.

Even critics of the Bush administration—those who emphasized the preservation of civil liberties and the restoration of American values—often overlooked the forms of unmarked racism and the new lines of social stratification upon which fragmented security ultimately now rests. Fragmented security expands the social and civic capacities of law enforcement and the criminal justice system more generally, giving new credence to thoroughly discredited tactics such as racial profiling and selective prosecution. So although the new security regime allowed law enforcement to read the e-mail messages and listen to the phone conversations of curious academics who Google Al Qaeda (and this is indeed something to worry about), it is clear that, without a significant overhaul, the techniques of the new security regime are applied selectively, and that this is bad news for vulnerable immigrant and minority populations. In other words, without major changes, homeland securitization will continue for the foreseeable future to work to justify and legitimate skewed domestic priorities that make many U.S. cities and border regions increasingly polarized and unsafe for all but the very rich who can afford their own private security systems.

Also crucial to this new security regime is the figure of the homeland citizen, actively engaged in counterterrorist surveillance activities in his or her own neighborhood and workplace. Yet the mobilization of homeland citizenship has its own dangers as well. What we have seen is the polarization of U.S. citizenship around issues of homeland security, whereby some groups are presumed to be virtuous homeland citizens and others (such as refugees and immigrants) are treated with considerably more suspicion. Moreover, the devolution of crime-fighting power and authority to citizen groups cannot be assumed to be an entirely positive development. Indeed, it is part of a wider process of security privatization that has given private security forces and corporations a greater role in law enforcement. It is reasonable to question some of these groups for operating outside the law in ways that endanger innocent people. Similarly, it is also fair to question the wisdom of citizen surveillance measures. Although there is a great deal to be said for the virtues of civic engagement and for broader civil society solutions to the nation's problems, it is also true that, historically speaking, many U.S. voluntary organizations have been virtuous toward their own at the expense of others, having long and sordid histories of exclusionary and discriminatory practices that are sometimes violently enforced. The empowerment of civic groups as part of the war on terror promises much of the same, with civic (and, in some cases, paramilitary) groups absorbing the responsibilities—and risks—of homeland securitization without being equipped to manage or overcome the potentially discriminatory effects of their own actions on the ground.

The post-9/11 marrying of homeland security and neoliberal economics introduced new risks to marginal and insecure urban and border populations. Consonant with historical patterns of marginalization based on class, race, and ethnicity, the homeland security matrix places undue burdens on marginal groups and threatens liberal and constitutional democratic rights at their core. As this chapter has argued, the new twenty-first-century U.S. homeland is fundamentally organized to acquire security for some at the expense of others. The unequal distribution of risk and responsibility implicit in the new U.S. security regime is so extreme that it poses a dire challenge to the foundational values of democratic culture, including social justice, civil rights, and structural protections for the marginalized. This is certainly not the first time in U.S. history that the overarching civic and governmental concern for the commonweal has been sacrificed in the name of security. Whether the Obama administration will address the inequalities produced by this now-entrenched system of fragmented security is an open question.

Notes

1. In considering the new scope of surveillance and policing strategies, we must, of course, be mindful of the more or less continual use and abuse of coercive state power, at all levels, throughout the history of the United States, including but certainly not limited to slavery, Native American genocide, Jim Crow, immigration restrictions, forced sterilization of poor women, the Palmer raids, Japanese internment camps, McCarthyism, COINTELPRO, the war on drugs, and the prison-industrial complex. The rise of the homeland security state represents the latest incarnation of a state that corrupts and vitiates the constitutional ideals upon which it was founded. It is our contention that the evident—and palpable—novelty of our 9/11 conjuncture must be analyzed dialectically, in a dynamic of sameness and difference from preexisting state formations.
2. The neoliberal pattern of coercive control and surveillance of certain segments of the U.S. population is already well known and well studied (Goode and Maskovsky 2001; Mullings 2003, 2006). A wave of mass incarceration since the 1970s has disproportionately targeted people of color and African Americans in particular. This has occurred parallel to dramatic shifts in U.S. labor markets that have frozen wages for a growing segment of working people and parallel as well to the precipitous withdrawal of publicly funded health and social services. In *Lockdown America* (2000), Christian Parenti views the rise of mass incarceration as the repressive arm of a wider, larger class war brought about primarily to prevent rebellion by the poor who have been harmed by the neoliberal withdrawal of welfare and wages. Angela Davis (1998) views it as a form of "masked racism" that has undergirded the rise of a new prison-industrial

complex, a system of crime control that has produced profits by the superexploitation of incarcerated people of color. Goode and Maskovsky (2001) argue that from the perspective of the poor, these changes can be seen as a shift from the welfare state to the law-and-order state.

3. Information on the Citizen Corps is available at http://www.citizencorps.gov/index.shtm (accessed February 15, 2007).
4. Ibid.
5. Ironically, the Bush administration's funding of the Urban Areas Security Initiative and of related initiatives such as the Metropolitan Medical Response System and the Citizen Corps program has been criticized from the start as inadequate. Moreover, counterterrorism funding to major metropolitan areas such as New York City was cut drastically in 2006, while funding for smaller, more remote cities such as Charlotte, North Carolina, was increased. New York City's mayor, Michael Bloomberg, joined Democrats in Congress in accusing the Bush administration of squandering homeland security funds with pork-barrel spending (Al Baker and Diane Cardwell, "City Has Itself to Blame for Terror Cuts, U.S. Says," *New York Times*, June 2, 2006). This is but one example of the contentious politics of risk assessment that has played out since the program's inception.
6. Security studies is now a burgeoning area of academic research. Paving the way, one early study attempted to assess the economic impact of antiterrorism securitization on urban business districts and commercial centers (Glaeser and Shapiro 2001).
7. The slogan "If you see something, say something" was first used in 1993 in an ad campaign by the Metropolitan Transportation Authority (MTA). In 2003, it revived the slogan in its new antiterrorism campaign, "The Eyes of New York." The MTA advertised that it received almost 2,000 tips on its antiterrorism hotline in 2006. Yet the effectiveness of the campaign is still unknown. A spokesperson from the MTA admitted during a press conference in the summer of 2007 that she did not know how many of the hotline calls were legitimate and how many were false alarms (Karni 2007).
8. These raids—and the state's intrusion into the private spaces of the home—have led many cities, such as Washington, D.C., New York, Los Angeles, Miami, Denver, Phoenix, and San Francisco, to adopt "sanctuary" ordinances that ban police from interrogating people about their immigration status. In 2007, accusations of favoring the so-called sanctuary cities started to surface in heated public debates among Republican contenders for the presidential nomination.

References

Andreas, Peter, and Thomas J. Biersteker. 2003. *The Rebordering of North America: Integration and Exclusion in a New Security Context.* New York: Routledge.

Beresford, Annette. 2001. "Homeland Security as an American Ideology: Implications for U.S. Policy and Action." *Journal of Homeland Security and Emergency Management* 1, no. 3. Available at http://www.bepress.com/jhsem/vol1/iss3/301.

Canada-U.S. Smart Border Declaration. 2001. "The Smart Border Declaration, Foreign Affairs, and International Trade Canada," December 12. Available at http://www.dfait-maeci.gc.ca/anti-terrorism/declaration-en.asp.

Davis, Mike. 1992. "Fortress LA: The Militarization of Urban Space." In Michael Sorkin, ed., *Variations on a Theme Park: The New American City and the End of Public Space.* New York: Hill and Wang, 154–171.

Euskirchen, Markus, Henrik Lebuhn, and Gene Ray. 2007. "From Borderline to Borderland: The Changing European Border Regime." *Monthly Review* (November). Available at https://www. monthlyreview.org/1107euskirchen-lebuhn-ray.htm.

Glaeser, Edward L., and Jesse M. Shapiro. 2001. "Cities and Warfare: The Impact of Terrorism on Urban Form." National Bureau of Economic Research Working Paper No. 8696. Available at http://ideas.repec.org/p/nbr/nberwo/8696.html.

Goode, Judith, and Jeff Maskovsky. 2001. *The New Poverty Studies: The Ethnography of Power, Politics, and Impoverished People in the United States*. New York: New York University Press.

Johnson, Chalmers. 2004. *Blowback: The Costs and Consequences of American Empire*. New York: Macmillan.

Karni, Annie. 2007. "The MTA Has a Message for Television Watchers." *New York Sun* (July 18). Available at http://www.nysun.com/new-york/mta-has-a-message-for-television-watchers/58616.

Maskovsky, Jeff. 2003. "Global Justice in the Postindustrial City: Urban Activism beyond the Global-Local Split." In Jane Schneider and Ida Susser, eds., *Wounded Cities: Destruction and Reconstruction in a Globalized World*. New York: Berg, 149–173.

Massey, Douglas A., and Nancy A. Denton. 1993. *American Apartheid: Segregation and the Making of the American Underclass*. Cambridge, MA: Harvard University Press.

Office of Housing and Community Development. 1998. *Neighborhood Transformations: The Implementation of Philadelphia's Community Development Policy*. Philadelphia, PA: Office of Housing and Community Development.

Parenti, Christian. 2000. *Lockdown America: Police and Prisons in the Age of Crisis*. London: Verso.

Pratt, Anna. 2005. *Securing Borders: Detention and Deportation in Canada*. Vancouver: University of British Colombia Press.

Sibley, David. 1995. *Geographies of Exclusion: Society and Difference in the West*. New York: Routledge.

Sugrue, Thomas J. 1998. *The Origins of the Urban Crisis: Race and Inequality in Postwar Detroit*. Princeton, NJ: Princeton University Press.

Wilson, William Julius. 1987. *The Truly Disadvantaged: The Inner City, the Underclass, and Public Policy*. Chicago: University of Chicago Press.

Section 2 THREATS

Chapter 5

The Challenge of Bio-Terrorism

By Daniel S. Hamilton and Bradley T. Smith

Bio-terrorism poses a unique, strategic threat to the transatlantic community. A bio-terrorist attack in Europe or North America is arguably more likely than and could be as consequential as a nuclear attack, but requires a different set of national and international responses. Current systems to manage national and international epidemics of infectious disease have been consistently stretched to their limits by SARS and other natural outbreaks, and are wholly inadequate for the unique challenges of bio-terrorism. Europeans and Americans alike are ill-prepared either to prevent bio-terrorist attacks or to mitigate their consequences. Building societal resilience to the threat of bio-terrorism requires political leaders and security experts to recognize that epidemics unleashed intentionally by a thinking enemy are significantly different from other security threats, and demand coordinated and complementary US and European efforts in prevention, preparedness and response.

This chapter begins by describing the unique challenges of bio-terrorism. It assesses the difficulties of response, distinguishing between bio-safety, non-proliferation, and bio-defense, and assesses current transatlantic and global efforts. It then suggests ways forward in each of these three areas.

Bio-terrorism: A Unique Strategic Threat

There are essentially two types of bio-weapon agents. The first are pathogens—disease-causing microorganisms such as bacteria and viruses (e.g. the causative agents of anthrax, smallpox, plague). The second are toxins–nonliving molecules that are produced by a plant, animal, or microorganism, but that do not replicate in the body (e.g. ricin, botulinum toxin).[1] Pathogenic microorganisms have the potential to be far more dangerous than toxins because many microbes cause contagious disease that can result in devastating epidemics (e.g. smallpox).

There are a number of reasons to consider pathogenic bio-terrorism, and the epidemics it may cause, as a unique strategic threat to the transatlantic community. Pathogenic bio-agents can be extremely lethal and are hard to detect. They are cheap, they are accessible, and they are easy to transport.

They can reproduce, and they have the potential to be made more powerful and diverse as bioscience rides the tides of globalization.

Only biological weapons rank with nuclear weapons in their lethality. The US Congressional Office of Technology Assessment (OTA) has estimated that an attack, under ideal conditions, with 100 kg of aerosolized *Bacillus anthracis* spores (i.e. the causative agent of anthrax) over an urban area could cause as many as three million casualties, rivaling the lethality of a thermonuclear weapon.[2]

It has been estimated that the 2 g of powder contained in the anonymous letter sent to Senator Thomas Daschle in October of 2001 contained over two trillion *Bacillus anthracis* spores.[3] Given the best medical estimates of the lethal dose of inhaled spores, these 2 g could contain thousands or even millions of lethal doses–assuming perfect distribution to each person and no medical intervention.[4] This is an extraordinary amount of lethal potential, considering that 2 g of this powder can fit in the palm

Figure 2.1 The lethality of biological weapons mirrors that of nuclear weapons (source: US Congress, Office of Technology Assessment, *Proliferation of Weapons of Mass Destruction: Assessing the Risks, OTA-ISC-559* (Washington, DC: US Government Printing Office, August 1993).

Note

Analyses by the US Congressional Office of Technology Assessment determined that the estimated number of deaths that could result from the detonation of a 1 megaton hydrogen bomb in a major US city would be similar to the deaths that would occur from an outdoor, line-source aerosol release of 100 kg of anthrax powder under ideal meteorological conditions. In this study, only biological weapons matched the lethal potential of nuclear weapons—not chemical weapons or conventional explosives. An indoor release of a biological weapon has the potential to be an even more efficient attack than on outdoor release; resulting in more lethality per kilogram—or even gram—of material.

of one's hand. A biological weapon could be far less concentrated than this and still have a devastating effect. Significantly, most of the people who became ill with inhalational anthrax (and four of the five who died) in October 2001 did not actually open a letter–their exposures were from indirect means (i.e. dust from mail sorting machines, or cross-contamination of their mail with one of the anthrax letters), and they did not receive prompt medical care.

Pathogenic organisms are cheap and accessible. Many are present naturally in the environment. Natural outbreaks of anthrax, plague, Ebola, and other dangerous pathogens occur in humans and animals throughout the world almost continuously.[5] In addition, there are thousands of biomedical laboratories, hospitals, biotechnology firms and universities worldwide, containing stocks of pathogens and the tools of molecular biology and bio-manufacturing required to engineer and produce wet or dry preparations of biological agents.[6] The knowledge and equipment required to grow pathogens, modify them, and turn them into weapons is essentially a matter of public information, available on the Internet. To test the feasibility of assembling a covert bio-terror lab, the US Government successfully built a "proof-of-concept" covert production facility in an abandoned building in Nevada for less than $1 million using a handful of personnel and commercially available equipment, illustrating that a large industrial infrastructure is not required for production of bio-weapons.[7] Additionally, many of the techniques required to "engineer" pathogens to render them potentially more dangerous are everyday skills of biologists throughout the world, and are used for entirely beneficent purposes.

Pathogens can reproduce. Unlike fissile materials, infectious microorganisms reproduce rapidly under the right conditions. For example, a small "seed culture" or even just one *Bacillus anthracis* bacterial cell could be cultivated under optimal growth conditions to trillions of organisms in a matter of days. For this reason, biological agents stored in a laboratory cannot be inventoried in the way fissile materials can. The theft of even minute quantities of a pathogen can pose a security threat. Moreover, trade in microbial cultures is poorly regulated within and among most nations, although this is beginning to change.[8]

Although skeptics charge that biological weapons are hard to produce–that weaponized *Bacillus anthracis*, for instance, is too hard to mill or too difficult to disperse–the technologies needed to build aerosolizable weapons are readily available and are being used every day for perfectly legitimate purposes such as pharmaceutical applications and agriculture. The science of aerosols has advanced greatly since the end of the offensive US bio-weapons program in the late 1960s. The delivery of medicines via inhalers has spawned significant research into how to aerosolize drugs, proteins, and organisms.[9] There is an entire industry focused on the aerosol dissemination of biological agents over large areas for control of pests in agricultural settings. One of the prime agents used for this agricultural application is *Bacillus thuringiensis*, a close relative of *Bacillus anthracis*. A recent study of the spraying of *Bacillus thuringiensis* near the city of Victoria in Canada found that such spraying results in significant amounts of biological agent being produced in a form and particle size-range that is easily inhaled into human lungs, where inhalational infections would begin.[10]

Biological weapons are diverse, and thus offer terrorists and military planners considerable flexibility. US experts were surprised to learn of some of the agents that Iraq and the former Soviet Union had chosen to produce and weaponize.[11] At least 30 known pathogens have the physical and biological characteristics needed for mass casualty–producing biological weapons. Most national biological warfare programs have focused on ten to 15 agents.[12] Even this shorter list of biological weapons, however, offers a range of possibilities, from the lethal *Bacillus anthracis* to incapacitating agents such as *Coxiella burnetii* (which causes Q fever) and Venezuelan equine encephalitis virus. Pathogens that cause contagious diseases that have been developed as biological weapons include smallpox virus and *Yersinia pestis* (the cause of plague). This list of agents, however, reflects only known threats. The continual emergence of new infectious diseases (e.g. Ebola hem-orrhagic fever, SARS) and the accelerating power of biotechnology and genetic engineering mean that unexpected or novel pathogens may also emerge as threats.

Stunning progress in biotechnology, advanced food processing and pharmaceuticals promise untold benefits for humankind. The deciphering of microbial genomes as well as the human genome has been a massive multiplier of the power of individual scientists and small groups to accelerate beneficent research. But this knowledge is dual-use. The same information can be used by a thinking enemy–whether a state, a group, or even an individual–to create antibiotic resistance, to avoid the human immune system, or even to engineer virulent new pathogenic organisms.[13] As the power of biology expands throughout the twenty-first century, the spectrum of possible threats will grow. Well-understood "twentieth century" bio-weapons such as *Bacillus anthracis* and smallpox virus are likely to be joined by these engineered, twenty-first century bio-threat agents.[14]

Bio-weapons are portable and reloadable.[15] They can be built and deployed by individuals or small groups using small-scale, dual-use equipment, with no discernable "signature" that an offensive program is underway. For example, in spite of the industrial scale of the former Soviet bio-weapons program, Western intelligence had not been able collect enough evidence of its existence to convince political leaders to take action until defectors came to the West in the 1990s. While fissile materials give off ionizing radiation that can be picked up by instruments at a distance, biological pathogens and toxins have no comparable signatures that can be detected at a distance with currently available technologies.[16] A terrorist could smuggle a few grams of a dry powder preparation of a bio-weapon in a sealed plastic vial through a security checkpoint with little risk of detection.[17] Bio-terror weapons can be deployed through unusual routes (such as the mail or low-tech sprayers) that make them hard to identify, treat and contain. Because bio-weapons are relatively easy to deploy, they are also relatively easy to reload–a thinking enemy could carry out covert "sneak" attacks on multiple cities or regions simultaneously or in quick succession. This "reload" potential could significantly enhance the lethality of possible attacks, confound possible countermeasures and response efforts, and further heighten fear and uncertainty among publics.[18]

A biological weapon is unlikely to come with a specific "return address," as does a ballistic missile or even a truck bomb. While there is the potential for microbial forensics experts to use "genetic fingerprints" to link bioweapons back to a lab or geographical region, this is very challenging. The required forensic technologies are just beginning to be developed.[19] Even after almost three years and one of the most intensive investigations in the history of the FBI, there have been no arrests in the 2001 anthrax letters case. The extreme difficulty in tracing a covert biological attack to any person or nation compromises what is perhaps the most effective tool nations have to deter attacks with weapons of mass destruction–the capacity for rapid attribution and retaliation. Deterrence theories rooted in the nuclear stand-off of the Cold War are unlikely to apply to bioweapons in an age of catastrophic terrorism.

A further distinguishing quality of bioterrorism is its delayed effect.Those attacked with bio-weapons may not even know they have been attacked.The most likely result of a covert bioterrorist attack is an epidemic that is not recognized until days after the actual attack takes place, when ill people begin to appear at hospitals, clinics and doctors' offices. Even then, proper diagnosis is likely to be challenging and time-consuming, since the earliest symptoms of such diseases may be non-specific and "flulike." And such an attack is unlikely to produce an obvious "ground zero" that draws fire fighters, police, and ambulances—or immediately galvanizes political leaders and their national security apparatus. Moreover, the explosive increase in flows of people, goods and ideas around the globe that has occurred in the last few decades would amplify the effects of delayed identification of the attack. What were once local curses now have global range. In 1918, it took six weeks for pandemic flu to circle the globe. Now someone who gets sick in Africa can be in Los Angeles the next day. In the SARS outbreak of 2003, one of the earliest patients traveled from Singapore to New York to Frankfurt before his symptoms became apparent.[20] Numerous exercises and simulations have demonstrated current US vulnerability to clandestine attacks with biological weapons, and these exercises have only brushed the surface of what would be involved in an international response to a bio-terror attack.[21]

Finally, this threat is both real and growing. The United States was attacked with anthrax in 2001. Radicals based in London and Lyon managed to manufacture and test toxins like ricin and botulism, presumably for attacks across Europe, before being arrested by the authorities. A dozen countries are estimated to either possess or be seeking an offensive bioweapon capacity–including all those named by the US Department of State as sponsors of terrorism.[22] A number of non-state actors have sought to develop mass casualty bioweapons. The Japanese apocalyptic cult Aum Shinriyko worked hard to develop anthrax and botulinum toxin bioweapons, but fortunately was unsuccessful.[23] Documents and computer hard drives seized during the March 2003 capture of Khalid Shaikh Mohammed, a key operational planner for al-Qaeda, revealed that the organization had recruited a Pakistani microbiologist, acquired materials to manufacture botulinum toxin, and developed a workable plan for *Bacillus anthracis* production.[24] The biological sophistication of al-Qaeda and other non-state terrorist groups is inexorably increasing.[25] During the Soviet period, Russian bioweapons scientists successfully weaponized smallpox, *Bacillus anthracis*, and other bioweapons on an industrial scale. Today, many of these scientists are languishing in a crumbling scientific system, are not being paid well, and are recruitment targets for other countries or terrorist networks.

The Challenge of Preventing and Responding to Bio-terrorism

The unique characteristics of bio-weapons enhance their appeal to those seeking to damage the US, Europe or other nations. Bio-weapons pose a strategic threat to the transatlantic community for which our vast experience with nuclear safeguards, security and deterrence unfortunately does not prepare us. Preventing and responding to bio-terrorism will require a new approach that advances efforts along three distinct, mutually reinforcing tracks: bio-safety, non-proliferation, and bio-defense.

Bio-safety involves measures to prevent the accidental release or theft of pathogens from laboratories and medical facilities. Non-proliferation involves measures that make it more difficult for nation-states, terrorists or criminals to obtain and use deadly pathogens and toxins by focusing on horizontal proliferation (i.e. efforts by non-state actors to acquire bio-agents) and vertical proliferation (i.e. efforts by states themselves to acquire such agents). Finally, bio-defense involves measures of medical and public health preparedness and emergency response to effectively counter a bio-attack or outbreak should either of these preventative measures fail.

Traditionally, non-proliferation has been an issue for the diplomatic community, while bio-safety and bio-defense have been the preserve of the medical, public health, and bioscience communities. (After September 11 and the subsequent anthrax attacks in the US, bio-safety and bio-defense efforts were redoubled.) In an age of catastrophic terrorism and with the potential power of biological weapons, however, true security will require a broad-based protection-in-depth that advances on all these fronts simultaneously, and engages these disparate communities together.[26]

Issues of bio-safety

To protect the health of laboratory workers and the public, the biomedical community has developed standards and procedures to minimize the risk of accidents. Work involving pathogens themselves or clinical samples from infected patients (e.g. blood, sputum) is guided by these standards, which define proper laboratory procedures, the design of bio-containment laboratories, the keeping of detailed records, and safe protocols for transporting potentially dangerous samples.

Since the anthrax attacks of 2001, policymakers in the US have focused more intently on bio-safety issues. As a result, the requirements to monitor laboratory stocks, physical security, and distribution of pathogens on the "select agent" list of potential bio-weapon pathogens maintained by the US Government have increased significantly.[27] Laboratory accidents do have the potential to trigger outbreaks with international implications, as was seen in late 2003 and early 2004 when three separate outbreaks of SARS were traced to such accidents in Singapore, Taiwan, and China.[28] However, there is also concern that regulations that are too restrictive can hamper research that is critical to finding new treatments and vaccines for these pathogens.[29]

France, Germany, Israel, Great Britain as well as the United States have all passed domestic legislation relating to laboratory security. Additionally, in June 2003 the Council of the European Union (EU) called on EU members to develop best-practice guidelines to promote the enactment of national legislation for controlling dangerous pathogens and toxins.[30] To be effective, however, these disparate national regulations must be comprehensive as well as equivalent, if not necessarily uniform. The alternative is a patchwork of inconsistent regulations, including pockets of weak implementation or enforcement that could result in accidental release of a pathogen or the possible theft of seed stocks by terrorists from poorly protected facilities.[31] Nevertheless, bio-safety regulations have their limitations–laboratories are not the only source of pathogens. Many potential bioweapons are present in nature (e.g. *Bacillus anthracis),* and new biotechnologies allow pathogens to be created "from scratch," without needing to possess the actual organism, using DNA synthesizing machines and other molecular biology techniques (as has been done, for example, with polio virus[32]). Therefore, these bio-safety precautions must be taken in concert with effective non-proliferation and bio-defense policies.

Issues of non-proliferation

Non-proliferation involves efforts on two fronts: (1) on-the-ground measures to make it difficult for terrorists or criminals to obtain deadly pathogens or toxins from laboratories, which is often termed "horizontal" proliferation; and (2) efforts to prevent states themselves from acquiring or developing bioweapon capabilities, which is often termed "vertical" proliferation. The latter also involves issues of arms control.

Efforts to block horizontal proliferation have been advanced primarily through better on-the-ground security measures. However, the fact that microbial pathogens exist naturally in the environment and that small samples (even a single organism) can be grown into large quantities means that biological organisms cannot be controlled as effectively, or in the same manner, as fissile materials. Since no technology or procedure can ensure the quantitative accounting of pathogens in a laboratory collection, applying the controls based on nuclear safeguards and auditing of weapon stockpiles to biological laboratories would create a false sense of security while seriously impeding legitimate research. It is therefore essential to develop a new security paradigm tailored to the unique characteristics of microorganisms and the vulnerabilities of the facilities that work with them.[33]

A major problem for non-proliferation efforts is the former Soviet bioweapons program. The Soviet empire not only left behind 30,000 nuclear warheads and enough highly enriched uranium and plutonium to make 60,000 more, together with 40,000 metric tons of chemical weapons; it also left behind the knowledge to produce and deploy industrial-scale quantities of an array of biological weapons including the smallpox virus.[34] Russia's dysfunctional economy and eroded security systems have undercut controls on these weapons, material and know-how, and increased the risk that they could flow to terrorist groups or hostile forces. It is a matter of clear self-interest for Europeans and North Americans to make sure that the members of the bioscience research community in the former Soviet Union (some of whom were former bioweaponeers) can earn a living and participate in the global scientific community, and not be tempted to join those who would use their talents against us.

To this end, the US and its partners have worked with Russia to develop a range of initiatives to deal with this challenge. The leading program is the Cooperative Threat Reduction (CTR) program, also known as the Nunn-Lugar program. The Nuclear Threat Initiative (a non-governmental organization) is also active, and an EU-Russia cooperative program for non-proliferation and disarmament exists as well. Together, these programs offer a variety of targeted efforts to improve the security of pathogenic collections held at former Soviet bio-weapons facilities, ensure safe storage and handling of pathogens at microbiological research centers, promote collaborative research designed to prevent proliferation of bio-weapon-related biotechnology, improve transparency, and enhance force protection capabilities.

Details regarding the full dimension of these programs are beyond the scope of this essay, but are readily available.[35]

Efforts to deal with the former Soviet Union's bio-security legacy have been uneven. At the June 2002 G8 Summit, the European Union, Canada, France, Germany, Italy, Japan, Russia, and the United Kingdom pledged to match US funding of non-proliferation programs–$10 billion over ten years. The G8 also announced the creation of a Global Partnership to confront the threat posed by the potential proliferation of nuclear, chemical, and biological weapons and materials emanating from the former Soviet Union. Since 2002, membership in the Partnership has been extended to a number of new donors. The Partnership's scope has also been extended to areas beyond the former Soviet Union. Unfortunately, even though leading European nations have now accepted the need to share the burden, commensurate with Europe's financial capacities, actual pledges made by the other G8 partners remain below their commitments, and few of the funds pledged have been allocated for actual Global Partnership projects. Moreover, these efforts remain heavily focused on nuclear non-proliferation, with some attention to chemical issues. Biological non-proliferation and bio-safety largely remain orphans of this effort. Except for a US commitment of $54 million for improved safety and security of dangerous pathogen collections in Russia, Kazakhstan, Uzbekistan, and Georgia, only France (€5 million) and Sweden ($130,000) have allocated any of their Partnership funds for work in the biological realm. The European Commission, Germany, the UK, Japan, Russia, Finland, the Netherlands, Norway and Switzerland have not allocated any of their Partnership funding to bio-safety or bioweapons non-proliferation issues.[36] Moreover, efforts in Russia have been hampered by the fact that while cooperation is ongoing in most facilities involved in the former Soviet biological weapons program, four former military facilities continue to refuse to cooperate with the Nunn-Lugar program.[37]

Another attempt to limit the proliferation of dual-use biotechnologies is the Australia Group, an informal consortium comprised of the US, most European nations, the European Commission, Argentina, Australia, New Zealand, Japan, and South Korea, which seeks to harmonize national export controls on dual-use chemical and biological materials and laboratory equipment. The group was established in 1985 following the widespread use of chemical weapons by Iraq during the Iran-Iraq war, but has since turned its attention to biological weapons as well. Significantly, it has also complemented its traditional focus on state-level proliferation with greater emphasis on non-state actors. The Australia Group's effectiveness is limited, however, due to the informal, non-binding nature of the organization's efforts.[38]

These operational efforts to enhance non-proliferation are complicated by the challenges facing biological arms control treaties. The crux of the problem is that it is very difficult to distinguish between offensive and defensive efforts. There are no internationally recognized boundaries between "offensive" and "defensive" biological research and development programs–essentially, all of biomedical research and laboratory equipment is dual-use. The only criterion is intent, which is an exceedingly difficult benchmark. For the past 30 years, the primary arms control focus of the international community has been on vertical proliferation. The centerpiece of such efforts is the 1972 Biological and Toxin Weapons Convention (BWC), signed by 145 nations, which bans the development, production, stockpiling, acquisition, and transfer of biological weapons.[39] As discussed above, the various attributes of bioweapons, however, make verification exceedingly difficult, and the BWC does not in fact include provisions for verification.

Efforts to craft a monitoring protocol were initiated by BWC members in 1994. These negotiations, which were slow and difficult, were halted in 2001 when the US Government rejected a draft protocol, introduced a series of proposals intended to serve as alternatives to a legally binding, multilateral accord, and called for an end to the negotiations. US officials declared the draft protocol to be a "blind alley," arguing that it was not intrusive enough to detect clandestine biological weapons activities yet was too invasive to safeguard proprietary or classified information held by the pharmaceutical companies or government labs that would be inspected under the proposed protocol.[40]

The US position has unleashed a vigorous debate and contributed to transatlantic dissonance on the issue. Key allies argued that despite the draft protocol's limitations, it was still a useful step forward. The British Government, for example, while also expressing its preference for stronger measures to ensure compliance and transparency, made the point clear:

> ... the Protocol would have delivered significant benefits for transparency, monitoring and deterrence in key dual-use areas capable of misuse ... It would as such help to deter and investigate suspected non-compliance, whether concerning the activity of a particular facility, an alleged use of biological weapons or a suspicious outbreak of disease.[41]

In short, despite a range of initiatives, the overall picture is of an international effort that is suffering from the fact that essentially all of biology is dual-use and, unlike the nuclear arena, cannot be divided into clear offensive and defensive activities. In spite of these challenges, non-proliferation efforts can play a role in increasing global security, but there is a growing gap between the progress that is being made and the scope and urgency of the bio-weapons threat.

Issues of bio-defense

In contrast to issues of bio-safety and non-proliferation, the international challenges of bio-defense–efforts by medical and public health professionals to prepare for and respond to a bio-weapon attack–have received comparatively little attention with respect to transatlantic collaboration. Yet, given the challenges of preventing a bio-attack, bio defense and crisis response efforts are critically important. While other global security threats (e.g. nuclear, chemical, radiological) have an immediate effect, and response efforts can only "pick up the pieces," the incubation period of most infectious diseases means that many of the people infected in the attack could be saved if an effective public health and medical response were to be mounted. It is even possible that the existence of a robust transatlantic (or international) response capability could serve to deter a bio-terrorist attack.

The ability to mount an effective bio-defense requires that government leaders have adequate situational awareness of the epidemic so that they can lead their publics and make good decisions. The leaders must also have at their disposal a robust medical response capability that will require both trained personnel and effective drugs, vaccines, and other medical counter-measures. These requirements are made even more challenging by the likely need to cooperate across borders to defend multiple national populations.[42]

Attaining situational awareness during an epidemic will be difficult given the pervasive uncertainties of size, scope, and timing that characterize an epidemic, especially one with the reload potential of a deliberate attack by a thinking enemy. In addition, in the transatlantic community, political leaders and even some national security and health officials are unfamiliar with bio-weapons and have limited direct experience with large epidemics of infectious disease–natural or intentional. A robust public health system is critical to attaining situational awareness of an ongoing outbreak, yet for years the nations of the transatlantic community have largely neglected their public health infrastructure. Even with the multibillion dollar investment in US public health since September 2001, significant shortfalls in bio-defense preparedness still remain due to decades-long neglect.[43] This uncertainty and lack of public health capacity will pose considerable challenges when international leaders are pressed to coordinate their actions.

Moreover, a bio-terrorist attack in Europe is likely to generate immediate tension among EU, national, and local authorities, given Europe's multi-jurisdictional setting.[44] Officials are likely to be tentative or uncertain in their initial response, and the lines of jurisdictional "competency" or authority are likely to be unclear, to differ, or to conflict. Similar issues are likely to arise between federal and state/provincial authorities within North America. The SARS epidemic in 2002–2003, for example, underscored that the EU is not institutionally prepared to deal with rapidly moving transnational diseases. While the EU and member states had a system in place to monitor the spread of the SARS virus, they had no system for advising–let alone deciding–on EU-wide measures to contain it.[45] The recent integration into the European Union and NATO of new, largely poorer member nations with uneven bio-terrorism preparedness capabilities could further enhance transatlantic vulnerability to infectious disease unless such integration incorporates more systematic pan-European and transatlantic efforts to enhance infectious disease and bio-terrorism preparedness coordination.[46]

The ability to mount an effective medical response to a bio-attack is severely limited due to the developed world's neglect of infectious disease and public health for several generations. Health care is treated increasingly as a business; disaster and epidemic preparedness has often been neglected as an expensive "luxury." The economics of health care and just-in-time efficiencies have eliminated all surge capacity from hospitals in the US, and the situation is very similar in Europe. Hospitals today are simply not prepared for an influx of several dozen, let alone several thousand, ill and potentially contagious patients who will need complex treatment in an intensive care unit setting.[47] During the SARS outbreak in Toronto, only a few hundred cases crippled the health-care system for the entire five-million-plus person metropolitan region, and the city had to send out an urgent call to Canada and the US for volunteer doctors and nurses to shore up the system.[48]

In addition to a lack of medical response capacity, there is a severe lack of the tools required by doctors and nurses to tend to the ill–drugs, vaccines, diagnostic tests, and other countermeasures–and there is almost nothing in the pharmaceutical development pipeline. Claire Fraser, a prominent biologist and President of the Institute for Genomic Research (TIGR) in the US, recently wrote:

> The decisions within the pharmaceutical industry to abandon infectious disease research programs, which reflect today's economic reality, are coming at a time when the specter of antibiotic-resistant organisms is growing ever larger ... Our arsenals against microbial foes are inadequately stocked, and we are vulnerable not only to deliberate attacks but to natural outbreaks of emerging and reemerging diseases.[49]

There are only five major vaccine manufacturers in the world,[50] and a recent study found that of the 506 new drugs currently under development by the 22 largest pharmaceutical and biotech companies, only six were antibiotics.[51] Developing a new drug takes 5–15 years and costs several hundred million dollars (one prominent estimate puts the price tag at over $800 million).[52] The cost of development coupled with small markets in the developed world have driven pharmaceutical companies away from treatments for infectious disease and towards blockbuster drugs against chronic diseases with multi-billion dollar annual sales. The US Government's BioShield, a mechanism enabling the government to purchase $6 billion of bio-defense countermeasures over ten years,[53] is an important first step. However, given the size of the drug industry (a half-trillion dollar global market[54]), even $6 billion will likely not be enough to radically redirect drug developers or fill the antibiotic, antiviral, and vaccine pipelines. In addition, money alone will not solve this problem; a coherent countermeasure strategy in which industry, academia, and government work together as partners will be key.

Challenges to Transatlantic Bio-defense

The challenges of effective medical and public health preparedness and response at the national level are magnified significantly when considering the international aspects of preventing or responding to a fast-moving epidemic caused by an intentional attack of a contagious pathogen by a thinking enemy. Infectious disease does not recognize national borders. The international response to such a crisis will be influenced in particular by three key issues: transparency of information, movement of people, and sharing of resources. In each area there is serious need for more effective intra-European and transatlantic coordination, as well as coordinated transatlantic pressure to advance broader efforts within international bodies.

Transparency of information

Maintaining continual, open flows of information between public health officials, community physicians, security officials, and political leaders to provide situational awareness during an outbreak response is difficult yet essential. Even in a single metropolitan area, situational awareness and uniform decision-making can be a challenge–as was seen during the 2001 anthrax attacks, when public health agencies from Washington DC, Virginia, and Maryland initially could not agree on recommendations for treatment of anthrax, even though many of the sick and potentially exposed worked in one jurisdiction and lived in another. The Virginia and Maryland authorities followed the federal guidance from the Centers for Disease Control (CDC), while Washington DC officials did not.[55] This problem could be magnified many times when multiple nations are involved, due to considerations of national pride, economics, and security. Yet transparency is critical for the epidemiological investigation required to map and control the outbreak. In addition, knowing what neighboring nations are planning to do, especially how they plan to care for their citizens (e.g. vaccination plans), will also be critical for leaders to maintain public trust and confidence. If neighboring nations or close allies were to take dramatically different actions in repose to an epidemic or bio-attack, in order to maintain public confidence national leaders should be able proactively to explain these differences (and not be forced to react to news reports of actions in other nations).

The 2003 SARS outbreak offers a sobering example of how nations may react to an emergent epidemic within their nation. Although effective communication across national borders is critical to managing an epidemic, Chinese officials did not disclose the growing outbreak to the World Health Organization (WHO) until 10 February 2003, three months after it began on 16 November 2002 in Guangdong province. As the SARS outbreak spread outwards from East Asia to Europe and the Americas, the WHO became more and more concerned about the lack of Chinese cooperation. Ultimately, the WHO broke with normal protocols requiring it to act only in concert with local authorities, and felt compelled to express unilaterally its "strong concern over inadequate reporting" of SARS cases by the Chinese Government.[56] Four days after that WHO announcement—a full five months after the outbreak began–China fully acknowledged the dimensions of the epidemic, announced a "war on SARS," and initiated aggressive action.[57] Unfortunately, the five-month delay contributed to the spread of SARS to dozens of countries, ultimately resulting in over 8,000 infections and approximately 800 deaths. In this case, unilateral action by the WHO may have effectively prompted a nation to act, but only after the epidemic had spread beyond national boundaries.

Subsequent epidemics have shown that other governments have not necessarily learned China's lesson—issues of economics, trade, or national pride can still impact leaders' judgments in this arena. It took weeks for Thailand, the world's fourth largest poultry exporter, to reveal that it had chicken flocks infected with avian flu in early 2004. The Thai Prime Minister was accused of attempting quietly to contain the outbreak so as to not harm the nation's exports.[58] In the end, as happened with SARS, this lack of transparency only worsened the situation.

Currently, nations have limited tools to guarantee that their neighbors are acting transparently. The WHO is severely restricted in its ability to act independently of a national government. Generally, it must be invited into a nation, and must gain permission to investigate affected regions. The SARS outbreak has prompted the WHO's governing body, the World Health Assembly, to increase the authority of the WHO to act more independently to respond to an epidemic—new International Health Regulatons to codify these authorities were approved in May 2005.[59]

Not only is the international community as a whole ill-equipped to deal with information transparency during a bio-attack or other mass epidemic, the nations of the Atlantic Alliance themselves are also poorly prepared to cope with this challenge. There are some rudimentary mechanisms in place. The Euro-Atlantic Disaster Response Coordination Centre (EADRCC) was created in 1998 as the focal point for coordinating disaster relief efforts of the 46 Euro-Atlantic Partnership Council (EAPC) nations, but these efforts are not integrated into broader alliance security considerations, and have been focused primarily on technological accidents or natural disasters (such as floods or fires) rather than terrorist attacks.[60] EU nations also operate a series of disease surveillance networks, each managed by a different member nation, but they are designed to monitor disease during "peacetime." The intense pressures that would arise during the response to a bio-terrorist attack would likely stress these systems far beyond what they were designed for. It is not clear that the EU is institutionally prepared to deal, in Europe, with the pressures of a transnational, rapidly moving disease epidemic caused by a bio-attack, much less to provide assistance to other regions, including the US.[61] Additionally, the pre-existing public heath and infectious disease communications networks that do exist in the EU operate primarily between technical experts in public health or disaster preparedness. There is little direct contact with senior decision-makers or political leaders. After a bio-terrorist attack, however, communications and decision-making would immediately become strategic and should involve senior-level political leaders in affected nations.

The likely unevenness of European and transatlantic response was highlighted in September 2003 by Exercise Global Mercury, an international "table-top" simulation conducted under the auspices of the G7's Global Health Security Action Group to test communications between national health agencies in the event of a smallpox outbreak. EU member states focused on implementation of their national plans, the international aspects of national measures were seldom discussed, and no EU-wide strategy emerged to combat the outbreak. The exercise also highlighted the lack of international considerations in most national smallpox plans, and the dif-faculties involved in international exchange of information during such an emergency.[62]

Movement of people

Movement of infected people following an attack with a contagious disease could amplify the attack's initial effect, and would complicate the ability to attain situational awareness of the outbreak. As noted above, one of the earliest SARS patients flew from Singapore to New York to Frankfurt before his symptoms became apparent.[63] Unfortunately, there is no good understanding of how restriction of movement across international borders could help (or hurt) the response to an international bio-terror attack, and no preexisting standards or procedures between nations for restricting movement of people or trade in such contingencies. Many overlapping treaties, such as WTO, NAFTA or bilateral arrangements, obscure the legal framework for action. The Schengen agreement, which provides for free movement of people among 15 European countries, is not clear whether its emergency clause permitting temporary closure of borders could be triggered by a public health crisis. There is even less clarity within Europe, across the Atlantic, and internationally on procedures to restart international trave and trade once borders have been closed or movement of people restricted.

Sharing of resources

Another great challenge for international bio-terrorism response is the issue of sharing limited medical resources between nations during a crisis.Therapeutic drugs and vaccines will become severely limited during the response to a bio-terrorist attack, and there are restricted pre-existing mechanisms to share these vital resources between nations. Even doctors and nurses can become limiting in a crisis, as was seen during the Toronto SARS outbreak in 2003, when Toronto was forced to put out a call to US physicians and nurses to travel to Canada and care for SARS patients.[64] Stocks of smallpox vaccine are national secrets in many countries, and so complete statistics are difficult to obtain. However, it appears that national efforts to stockpile vaccine are very uneven. Information compiled by the European Commission in 2003 showed that national stockpiles of smallpox vaccine ranged from having one dose per citizen to having enough for one dose per 30 citizens.[65] It is likely that only a handful of nations in the world (US, UK, France, Germany, and perhaps a few others) have enough smallpox vaccine for their entire population. Many would be dependent on assistance from other countries–but which countries should receive such resources, and based on what criteria (e.g. close allies, a first-come-first-served basis, or every government that shows a need and asks for it)? It would be very difficult for any national leader to send precious vaccine to other nations when his or her own citizens were at risk. Discussions regarding developing a common European stockpile and having a sharing scheme of antibiotics and vaccines failed to lead to substantive action. Most member states, in fact, did not wish to continue these sensitive discussions, as any reform would require significant financial resources for what is still regarded as a low-probability problem.[66] The WHO does maintain a small stockpile of smallpox vaccine left over from the eradication campaign that could be used to combat a smallpox outbreak, and it is working to develop a "virtual" stockpile of vaccine from national stock-piles.[67] Plans for distributing this critical resource in the wake of a smallpox attack, however, are still being developed.

Moving Forward

Bio-terrorism poses a daunting series of interrelated security challenges for the transatlantic community. As Europe and North America work to lead an international effort to thwart this threat, primary consideration should be given to the following two factors.

First, there is a window of opportunity, but it is closing. Despite the vast potential for bio-weapons proliferation, thus far the spread of such weapons among states and non-state actors appears to have been limited. The power of biotechnology to do good or ill is increasing at an accelerating rate, and as a result we are on the cusp of exponential change, both in the power of bio-weapons as well as their accessibility to state and non-state actors alike. Effective action now could mitigate the negative impact of this technological trend.

Second, something can be done after an attack to minimize, or potentially eliminate, death and suffering. Drugs, vaccines and other medical counter-measures can save lives if there is a robust medical response capacity. The transatlantic community contains the highest concentration of biomedical talent in the world, as well as the bulk of the global pharmaceutical industry. Concerted efforts to build on this capacity and integrate it with robust governmental decision-making systems will save lives.

With these factors in mind, it is incumbent upon the US and Europe to build on their respective efforts to thwart bio-terrorism by working more effectively across the Atlantic. Efforts to bolster trans-European or transatlantic bio-terrorism preparedness will remain limited until medical, public health, and local civil emergency authorities are more effectively woven into the European integration process and transatlantic civil/homeland defense. High-level political awareness is critical, and commensurate political commitment and monetary resources are required. As former US Senator Sam Nunn has stated:

> We have to realize that we have reached a new realm in the dialectic of new weapons and new defenses. In the evolution of warfare, arrows were countered by shields, swords with armor, guns with tanks, and now biological weapons must be countered with medicines, vaccines and surveillance systems.

The US and Europe should advance a broad-based effort at protection in depth that includes mutually reinforcing initiatives in the areas of bio-safety, non-proliferation, and bio-defense.

New steps in bio-safety and non-proliferation

As discussed above, the almost universal "dual use" nature of biological sciences and the prevalence of many dangerous pathogens in nature limit the ability of bio-safety and non-proliferation efforts to prevent a biological attack. Nevertheless, such efforts have an important role to play as the transatlantic community seeks to build a layered defense against this strategic threat, and as cooperative international efforts to develop uniform standards and procedures are underway.

Bio-safety and non-proliferation could both be served by a more integrated global bioscience research community that develops shared standards and norms of behavior. The dual-use nature of bioscience and the inability to draw clear lines around "dangerous" science has led some to suggest that the scientific community should use a self-governance system to manage the gray area of dual-use science, and minimize accidental development of powerful bio-weapons.[68] However, others feel that a more formal regime is necessary and possible.[69] A recent National Academies of Science report recommended the more flexible self-governance model, and proposed that, in the US, a panel of scientists should evaluate potentially dangerous lines of experimentation and determine if the potential benefits outweigh the costs.[70] The US Government quickly accepted many of the report's recommendations, and established the National Science Advisory Board for Biosecurity (NSABB).[71] The NSABB is still in its planning stages, so it is unclear how the board will carry out its mission, and how effective it will be. Preexisting efforts that sought to engage scientists in the former Soviet Union (as the Nunn–Lugar CTR programs do) could be expanded and linked with these new calls for the biomedical research community in the US and Europe to consider the dual-use implications of its research. This could create a more unified scientific community, including researchers in the Americas, Europe, the former Soviet Union, and Asia, that is actively working to prevent proliferation of biological weapons.

The Organization for Economic Cooperation and Development (OECD), a group of 31 advanced industrial countries headquartered in Paris, is working to establish a global network of Biological Resource Centers (BRCs) that would act as a virtual lending library of microbial cultures based on harmonized, accredited standards and regulations for bio-safety. These arrangements will be based on voluntary agreement, but those countries failing to meet the standards will be excluded.[72] Efforts by the WHO and professional societies to improve and harmonize national standards in the wake of the recent laboratory accidents with SARS virus should be encouraged.[73]

Other international organizations have also entered the non-proliferation arena. The OSCE has proposed licensing standards and enforcement procedures related to dangerous pathogens and dual-use biotechnology equipment. The World Customs Organization has begun informal information-sharing with Interpol and the World Health Organization to combat the smuggling of biological, chemical, and radioactive materials. The International Maritime Organization plans to negotiate agreements to halt the shipping of biological agents for hostile purposes, and to criminalize the use of biological weapons on maritime vessels.[74] Additionally, as Jonathan Winer discusses in Chapter 6, the G7 responded to the danger of terrorist financing during the 1990s by launching a Financial Action Task Force to set global standards in this area. The FATF provides an interesting model of a networked response to a networked threat, and may offer some lessons learned for networked cooperation in other areas, including bio-weapon non-proliferation efforts involving non-state actors

Standard-setting can also help to invigorate efforts to counter vertical proliferation (i.e. state-sponsored programs). Many diplomatic initiatives have faltered due to failure to articulate an international standard that governments would be expected to meet. Absent identification of and agreement on such standards, governments have little to compel them to take action. Many governments will enact measures that fall short of worthwhile standards either unintentionally, because they cannot decipher the existing regulatory concepts, or intentionally, because they seek to perpetuate illicit activities. The let-each-government-do-as-it-pleases approach will further foster an uneven patchwork of domestic laws and practices that might have little near-term value and could prove difficult to harmonize in the future. Extending international standards and making them decipherable for translation in to national legislation will raise the bar.[75]

Making progress on transatlantic bio-defense

As discussed above, an effective bio-defense must provide situational awareness to decision-makers, and a robust medical response capacity. Although many of the medical and public health preparedness components of an effective bio-defense are nationally focused, infectious disease epidemics do not recognize national boundaries. International bio-defense cooperation–beginning with the nations of the Atlantic Alliance–will be critical to the success of any individual national effort. Key will be closer intra-European cooperation, together with new forms of cooperation between the US and the EU, and between NATO and its partners as well.

A potentially significant EU initiative is the creation of the European Center for Disease Prevention and Control (ECDC) in Sweden, to be operational in 2005. Its mission encompasses elements of bio-safety, non-proliferation, and bio-defense. The ECDC will operate as a network rather than a centralized agency, maintaining only a small office in Sweden. It will run no independent labs, nor will it consolidate any member state disease-control agencies. It will work as a coordinating entity in charge of pooling European resources and reinforcing those existing public health organizations across the European Union that are in charge of controlling communicable diseases.[76] Among other responsibilities, it will provide scientific advice to the EU Commission and national authorities, facilitate epidemiological surveillance and laboratory networking, operate the EU's early warning and response system (EWRS), and support the development of EU-level preparedness planning for health crises "such as an influenza pandemic or a bio-terrorist attack."[77] Although the ECDC has been envisioned as pursuing the same goals as the US Centers for Disease Control and Prevention based in Atlanta, Georgia, and is likely to become a focal point for transatlantic communication on disease surveillance and bio-defense issues, the ECDC will have very different operational capabilities and administrative structures.[78] Most importantly, the prerogative of deciding on a specific course of action in a crisis will still be in the hands of the member states. Nevertheless, the ECDC could serve as an important European partner for the US CDC and other nations if the ECDC is involved a coherent EU-wide decision-making process.

A second, related effort is the Global Health Security Action Group established by the health ministers of the G7 plus Mexico on 7 November 2001 in Ottawa. The group has worked on improving international communications and assistance capabilities in a health crisis, improving non-proliferation protocols for the transfer of pathogens from one laboratory to another, and promoting best practice in such areas as vaccine production and isolation techniques. The Global Health Security Action Group might consider extending its current work, which has been limited primarily to assessing communications in a crisis, to standard setting and medical resource-sharing. It could even engage G7 Interior and Foreign Ministers (in the US case, the Secretaries of Homeland Security and State) to consider the difficult issue of movement of people.

As NATO transforms itself for the twenty-first century, the Alliance's civil defense capabilities should be revamped and given much higher profile with respect to bio-terrorism preparedness and response. The Partnership for Peace and the Euro-Atlantic Partnership Council have undertaken several initiatives in the biological protection field. Potentially, the Euro-Atlantic Disaster Response Coordination Center (EADRCC) could be used as a focal point for an operative effort.

The EU has been working to improve its preparedness. The anthrax attacks in the US, together with uncovered plots to unleash toxins and pathogens in Europe, have combined with the March 11 Madrid attacks to prompt EU member states and the European Commission to take several interrelated initiatives to enhance bio-preparedness and response capacity. For example, they have created a bio-terrorism task force,[79] have reinforced mechanisms to share information (e.g. early warning, alert, notification), established detection and classification methods, launched a mechanism for acquiring information on the capacities of laboratory facilities in the member states (as no single state has the laboratory capabilities and resources needed), and assessed the availability of serums and vaccines in the member states.[80] Such efforts are slow, however, and do not match the urgency of the challenge. Moreover, while the EU slowly enhances its operational and technical preparedness, top-level decision-makers and national leaders in Europe have only begun to consider how they might act if they were thrust unexpectedly into an acute crisis-management mode in response to a sudden outbreak of a contagious disease within Europe or in the transatlantic community–something which, unfortunately, could happen today.

Bio-defense is dependent on having drugs, vaccines, and other medical countermeasures available for use by medical and public health officials during a crisis, and, as has been discussed above, there is a critical lack of such medicines today. The transatlantic community possesses the bulk of the world's academic and industrial biomedical research community. Integrating these scientists and drug development experts into a coherent, targeted transatlantic effort to develop new bio-defense and infectious disease countermeasures would be a powerful tool to lower the threat of bio-terrorist attack. Such an effort should have a long-term goal of accelerating the drug development process overall and developing new countermeasures that can contribute to the elimination of the lethal effects of epidemics caused by bio-weapons.[81] The bioscience community is inherently international. A recent study has found that cross-border collaborations in the biosciences are growing, and are important to the development of new medicines. The authors of the study also recommended that national bio-defense efforts should encourage such international collaboration, not discourage it for fear of aiding proliferation.[82] The US Congress has tasked the National Institutes of Health (NIH) with the development of new bio-defense countermeasures, and has provided the agency with approximately $3.3 billion for intra- and extramural bio-defense research funding since 2001, with another $1.7 billion expected in 2005.[83] Unfortunately, this scale of investment in bio-defense research has not been mirrored in Europe.

Accomplishing the goal of taking infectious diseases and bio-terror agents "off the table" will require enormous creativity, organization, and investment. Leaders on both sides of the Atlantic should strive to link US Government research efforts in infectious disease and bio-defense with Europe's, perhaps through the EU's "framework program" of scientific funding. Such a linkage, combined with a robust strategy that adequately engages and directs academic, industrial, and government researchers, could result in the development of the medicines needed not only to reduce the threat of bio-terrorism in the transatlantic community, but perhaps also to eliminate the scourge of infectious disease worldwide.

Conclusion

It is perhaps impossible to protect fully against any biological attack. The goal must be to enhance the resilience of European and North American societies to large-scale biological attacks so that it is not worth mounting them. There is an opportunity for transatlantic leadership on this issue, but time is not on our side. It is urgent that Europeans and Americans address the threat now.

Issues of public health must become mainstream issues of diplomacy and national security. This means offering opportunities in government for experts in these matters–the interface between governments and biomedical experts in academia and industry must become tighter. We will not succeed without a critical mass in governments of those who understand the science and can understand the threat. Europe and America have phenomenal capability in bioscience, medicine, and public health. Whether it can be mobilized and directed quickly enough to prevent or minimize the effects of a big bio-terrorist attack is an open question. Can we switch the advantage to the defense? Absolutely. But it is a matter of will and imagination.

Notes

1. See G. Koblentz, "Pathogens as weapons: the international security implications of biological warfare," *International Security,* 28(3), Winter 2003/2004. For technical background on biological weapons, see Office of Technology Assessment (OTA), *Proliferation of Weapons of Mass Destruction: Assessing the Risks,* Washington, DC: US Government Printing Office, 1993, pp. 71–117.
2. Office of Technology Assessment, op. cit., pp. 53–54.
3. B. Broad, "A nation challenged: the spores; terror anthrax resembles type made by US," *New York Times,* 3 December 2001.
4. T.V. Inglesby, T. O'Toole, D. A. Henderson *et al.,* "Anthrax as a biological weapon, 2002," *Journal of the American Medical Association,* 287, 2236–2252.
5. This is unlike plutonium and highly-enriched uranium, the essential ingredients for nuclear weapons, which are synthetic materials that do not exist in nature and are difficult and costly to produce.
6. Koblentz, op. cit. This is in contrast to about a hundred restricted nuclear sites around the world containing weapons-grade fissile materials.
7. J. Miller, "Next to Old Rec Hall, a 'germ-making plant'," *New York Times,* 4 September 2001.
8. This stands in stark contrast to the elaborate efforts in place to track and control the materials needed to build nuclear weapons. For further details see Koblentz, op. cit., and J. B. Tucker, *Biosecurity: Limiting Terrorist Access to Deadly Pathogens,* Washington, DC: US Institute of Peace, November 2003.
9. 9 D.A. Edwards and C. Dunbar, "Bioengineering of therapeutic aerosols," *Annual Review of Biomedical Engineering,* 4, 2002, 93–107.
10. D. B. Levin and G. V. De Amorim, "Potential for aerosol dissemination of biological weapons: lessons from biological control of insects," *Biosecurity and Bioterrorism,* 1(1), 2003, 37–42.
11. Tucker, op. cit.; K. Alibek and S. Handelman, *Biohazard,* New York: Random House, 1999.
12. Tucker, op. cit.; Koblentz, op. cit.
13. See G. Kwik J. Fitzgerald, T.V. Inglesby, and T. O'Toole, "Biosecurity: responsible stewardship of bioscience in an age of catastrophic terrorism," *Biose-curity and Bioterrorism,* 1(1), 2003, 27–35.
14. J. B. Petro, T. R. Plasse and J.A. McNulty, *Biosecurity and Bioterrorism,* Vol. 1(1), 161–168.
15. R. Danzig, *Catastrophic Bioterrorism—What Is To Be Done?* Washington, DC: Center for Technology and National Security Policy, National Defense University, August 2003.
16. A January 2003 workshop hosted by the CIA and the National Research Council brought together bio-scientists and members of the intelligence community to discuss the signatures (or "observables") that would indicate the presence

of an offensive bio-weapons program using sophisticated biotechnologies. The bio-scientists indicated in the strongest terms that there would be no such signatures–biotechnologies are all dual-use. An official summary of this meeting is available online at www.fas.org/irp/cia/product/bw1103.pdf (accessed 22 September 2004).

17. Tucker, op. cit.
18. R. Danzig, op. cit.
19. R. Murch, *Biosecurity and Bioterrorism,* 1, 2003 117–122. B. Budowle, S. E. Schutzer, A. Einseln *et al.*, "Building microbial forensics as a response to bioterrorism," *Science,* 26 September 2003, 1852–1853.
20. WHO update 95-SARS: Chronology of a Serial Killer, available online at: www.who.int/csr/don/2003_07_04/en/ (accessed 20 September 2004).
21. To begin to address this gap, a table-top exercise called *Atlantic Storm* was convened 14 January 2005 by the Center for Biosecurity of UPMC, the Center for Transatlantic Relations of Johns Hopkins University, and the Transatlantic Biosecurity Network. *Atlantic Storm* simulated a summit of transatlantic leaders forced to respond to bioterrorist attacks on multiple nations. The exercise focused on the high-level strategic, political, and security challenges that heads of government will be confronted with as they respond to the unique security threat of bio-terrorism. Further information on the exercise is available online at http://www.upmc-biosecurity.org./ and http://www.transatlantic.sais-jhu.edu See also J. Miller, "Exercise finds US unable to handle germ war threat," *New York Times,* 26 April 1998, p. A1; T. V. Inglesby, R. Grossman and T. O'Toole, "A plague on your city: observations from TOPOFF," *Clinical Infectious Diseases,* 32(3) 2001, 436–445; and T. O'Toole, M. Mair, and T. V. Inglesby, "Shining light on dark winter: lessons Learned," *Clinical Infectious Diseases,* 34(7), 2002, 972–983.
22. According to the US Government, roughly a dozen countries, including Iran and North Korea, are thought to have offensive bio-warfare efforts underway. US Department of State, International Information Programs, "Statement of Under Secretary of State for Arms Control and International Security John Bolton to the Fifth Review Conference of the Biological Weapons Convention," Geneva, Switzerland, 19 November 2001, available online at: www.state.gov/ t/us/rm/janjuly/6231.htm.
23. D. E. Kaplan, "Aum Shinriyko," in J. Tucker (ed.), *Toxic Terror,* Cambridge: MIT Press, 2000.
24. Cited in Tucker, *Biosecurity*, op. cit; B. Gellman, "Al Qaeda near biological, chemical arms production," *Washington Post,* 23 March 2003, p. A1. Case studies of terrorist groups interested in biological weapons can be found in J. B. Tucker (ed.), *Toxic Terror,* op. cit., and W. S. Carus, *Bioterrorism and Biocrimes: The Illicit Use of Biological Agents in the 20th Century,* Washington, DC: National Defense University, April 2001.
25. J. B. Petro and D. Relman, "Understanding threats to scientific openness," *Science,* 12 December 2003, 1898.
26. For additional perspectives, see M. L. Ostfield, "Bioterrorism as a foreign policy issue," *SAIS Review,* 24(1) 2004, 131–146.
27. Available online at www.cdc.gov/od/sap/ (accessed 21 September 2004); D. Malakoff, "Congress adopts tough rules for labs," *Science,* 31 May 2002, 1585–1587.
28. D. Normile and G. Vogel, "Early indications point to lab infection in new SARS case," *Science,* 19 September 2003, 1642–1643; D. Normile "Second lab accident fuels fears about SARS," *Science,* 2 January 2004, 26; D. Normile, "Mounting lab accidents raise SARS fears," *Science,* 30 April 2004, 659–660.
29. *Biotechnology Research in an Age of Terrorism,* Committee on Research Standards and Practices to Prevent the Destructive Application of Biotechnology, National Research Council, 2004; D. Malakoff and M. Enserink, "Biodefense: researchers await government response to self-regulation plea," *Science,* 17 October 2003, 368–369.

30. *House of Cards: The Pivotal Importance of a Technically Sound BWC Monitoring Protocol,* Washington, DC: Stimson Center, 2001.
31. Tucker, *Biosecurity,* op. cit.
32. J. Cello, A. V. Paul and E. Wimmer, "Chemical synthesis of poliovirus cDNA: generation of infectious virus in the absence of natural template," *Science,* 9 August 2002, 1016–1018.
33. J. D. Steinbruner and E. D. Harris, "Controlling dangerous pathogens," *Issues in Science & Technology,* 19(3) 2003; 47–54; G. Kwik, J. Fitzgerald, T. Inglesby, and T. O'Toole, "Biosecurity: responsible stewardship of bioscience in an age of catastrophic terrorism," *Biosecurity and Bioterrorism,* 1(1), 2003, 27–35; *Biotechnology Research in an Age of Terrorism,* op cit.
34. K. Alibek and S. Handelman, *Biohazard,* New York: Random House, 1999.
35. Joint US-Russian research is being conducted at 49 former biological weapons production facilities, and security improvements are underway at four biological weapons sites. See the statement by Senator Richard Lugar at the National Press Club, 11 August 2004, available online at: www.lugar.senate.gov/nunnlugar.html. The suggestion has also been made to expand European involvement in these programs, perhaps by expanding the scope of the EU's Joint Action plan toward Russia as the basis of an EU equivalent to CTR and a mechanism for coordinating bilateral arrangements between EU states and third countries. See D. Feakes, "The emerging European disarmament and non-proliferation agenda on chemical and biological weapons," *Disarmament Diplomacy,* 65, 2002.
36. For a detailed official outline of funding commitments and projects by country, see G8 Consolidated Report of Global Partnership Projects, available online at: www.g8usa.gov/pdfs/GPConsolidatedReportofGPProjectsJune2004.pdf. A related initiative is the Proliferation Security Initiative (PSI), a multilateral effort established in 2003 to interdict weapons and materials of mass destruction. More than 60 nations, including Russia, have associated themselves with this effort, but it too focuses primarily on nuclear issues. See the EU country report in Volume 3, *International Responses,* of R. Einhorn and M. Flournoy, *Protecting Against the Spread of Nuclear, Biological and Chemical Weapons,* Washington, DC: Center for Strategic and International Studies, 2003.
37. For an independent assessment of the Global Partnership's progress, see the CSIS monitoring effort, available online at: www.sgpproject.org/GP%20Score-card.pdf.
38. Available online, at www.australiagroup.net/ (accessed 14 September 2004); Koblentz, op. cit.; Tucker, *Biosecurity,* op. cit.
39. Available online at: www.opbw.org/ (accessed 20 September 2004); www.state. gov/t/ac/trt/4718.htm (accessed 20 September 2004).
40. "The draft protocol that was under negotiation for the past seven years is dead in our view. Dead, and it is not going to be resurrected. It has proven to be a blind alley." US Department of State, International Information Programs, "Bolton briefing on the biological weapons pact," transcript of press conference, 20 November 2001, available online at: www. usinfo.state.gov/topical/pol/terror/01112003.htm. See also Ambassador Donald Mahley, "Statement by the United States to the Ad Hoc Group of Biological Weapons Convention States Parties," Geneva, Switzerland, 25 July 2001, available online at: www.state.gov/t/ac/rls/rm/2001/5497.htm For a review of US participation in the negotiations over a verification protocol to the BWC, see M. Leitenberg, *Biological Weapons and "Bioterrorism" in the First Years of the 21st Century,* Center for International and Security Studies, School of Public Affairs, University of Maryland. Paper prepared for conference on "The Possible Use of Biological Weapons by Terrorist Groups: Scientific, Legal and International Implications," ICGEB, Landau Network, Ministry of Foreign Affairs, Rome, April 16, 2002, paper updated to 15 May 2002.
41. See *Strengthening the Biological and Toxin Weapons Convention: Countering the Threat from Biological Weapons,* UK Foreign and Commonwealth Office.

42. Koblentz, op. cit.; Tucker, *Biosecurity*, op. cit.
43. *Ready or Not? Protecting a Public's Health in an Age of Bioterrorism*, Washington, DC: Trust for America's Health, December 2003.
44. B. Sundelius and J. Gronvall, "Strategic dilemmas of biosecurity in the European Union," *Biosecurity and Bioterrorism*, 2, (1), 2004, 17–23.
45. EU Health and Consumer Protection Commissioner David Byrne noted that SARS was a "wake-up call" for Europe to get better prepared and to substantially enforce cooperation at an EU level.
46. R. J. Coker, R. A. Atun, and M. McKee, *The Lancet*, 363, 2004, 1389–1392.
47. The mean occupancy rate of US intensive care units (where many of the seriously ill victims of a bio-attack will need to be treated) in the largest US hospitals has been estimated to be 93 per cent. J. S. Groeger, K. K. Guntupalli, M. Strasberg *et al.*, *Critical Care Medicine*, 21, 1993, 279–291.
48. L. K. Altmann, "The doctor's world; behind the mask, the fear of SARS," *New York Times*, 24 June 2003.
49. C. M. Fraser, "An uncertain call to arms," *Science*, 16 April 2004, 359.
50. They are Merck and Co., Aventis Pasteur, GlaxoSmithKlein, Wyeth, and Chiron.
51. B. Spellberg, J. H. Powers, E. P. Brass *et al., Clinical Infectious Diseases*, 38, 2004, 1279–1286.
52. J. A. DiMasi, R. W. Hansen, and H. G. Grabowski, *Journal of Health Economics*, 22, 2003, 151–185.
53. US Public Law 108–276.
54. IMS *World Review*, 2004.
55. E. Gursky, T. V. Inglesby and T. O'Toole, *Biosecurity and Bioterrorism*, 1, 2003, 97–110.
56. "Learning for SARS," Washington, DC: Institute of Medicine, 2004.
57. "Learning for SARS," Washington, DC: Institute of Medicine, 2004.
58. A. Sipress, "Thailand admits mistakes in handling of bird flu outbreak," *Washington Post*, 28 January 2004.
59. C. H. Conde, "WHO seeks to change rules on contagious diseases," *International Herald Tribune*, 28 April 2004.
60. There have been some limited efforts to shift the EADRCC from its focus on natural disasters to premeditated attacks. The field exercise "Dacia 2003" in October 2003, proposed by Romania, focused on consequence management following a terrorist attack using a Radiological Dispersion Device ("dirty bomb"). Unfortunately, this remains a rather isolated example. For more information see "Dacia 2003", available online at: www.nato.int/eadrcc/2003/dacia/index.htm.
61. Sundelius and Gronvall, op. cit
62. S. Vedantam, "WHO assails wealthy nations on bioterror," *Washington Post*, 5 November 2003. For the full report, see *Global Mercury Report*, available online at: www.doh.gov.uk/cmo/exerciseglobalmercury/globalmercuryreport.PDF.
63. Available online, at www.who.int/csr/don/2003_07_04/en/ (accessed 20 September 2004).
64. L. K Altmann, "Behind the mask, the fear of SARS," *The New York Times*, 24 June 2003.
65. See Sundelius and Gronvall, op. cit.; Tucker, op. cit; Tegnell, op. cit.; Leiten-berg, op. cit.
66. Ibid.
67. The WHO currently has several hundred thousand doses of vaccine, but studies have shown that those doses likely could be diluted 1: 5 without losing efficacy. Available online, at www.who.int/csr/disease/smallpox/preparedness/en/ (accessed 20 September 2004).
68. G. Kwik *et al*, op. cit.
69. J. D. Steinbruner and E. D. Harris, op. cit.
70. *Biotechnology Research in an Age of Terrorism*, op cit. D. Malakoff and M. Enserink, op cit.

71. Available online, at www.biosecurityboard.gov/index.htm (accessed 20 September 2004).
72. See the OECD report, *Biological Resource Centers: Underpinning the Future of Life Sciences and Biotechnology.* Online, available at: www.oecd.org/dataoecd/26/ 19/31685725.pdf.
73. Available online at: www.who.int/csr/labepidemiology/projects/biosafetymain/en/ (accessed 14 September 2004); D. Normile, "SARS experts want labs to improve safety practices," *Science*, 3 October 2003, 31.
74. For a review of various agreements, see Tucker, *Biosecurity*, op. cit., and Koblentz, op. cit.
75. Tucker, *Biosecurity*, op. cit.
76. The agency will likely start in 2005 with a core staff of as few as 40 people and then grow to a full force of 100 people by 2007. See N. Stafford, "European CDC closer to reality," *The Scientist* website, available online at: www.biomed-central.com/news/20031204/04/; A. Tegnell, P. Bossi, A. Baka, F. Van Loock, J. Hendriks, S. Wallyn, and G. Gouvras, "The European Commission's Task Force on Bioterrorism," *Emerging Infectious Diseases*, October 2003. Online, available at: www.cdc.gov/ncidod/EID/vol9no10/03-0368.htm. See also "European Centre for Disease Prevention and Control (ECDC)," European Union website, available online at: europa.eu.int/comm/health/ph_overview/strategy/ecdc/ecdc_en. htm.
77. "European Centre for Disease Prevention and Control (ECDC)," European Union website, available online at: europa .eu.int/comm/health/ph_overview/ strategy/ecdc/ecdc_en.htm.
78. Stafford, op. cit.; Tegnell, op. cit.; "European Centre for Disease Prevention and Control Proposed," European Public Health Alliance website, available online at: www.epha.org/a/585.
79. A. Tegnell, P. Bossi, A. Baka *et al.*, *Emerging Infectious Diseases*, 9, 2003, 1330–1332.
80. Available online, at http://europa.eu.int/comm/health/ph_threats/Bioterror-isme/bioterrorisme_en.htm (accessed 20 September 2004). "European Centre for Disease Prevention and Control Proposed," European Public Health Alliance website. Online, available at: www.epha.org/a/585.
81. B. T. Smith, T. V. Inglesby, and T. O' Toole, *Biosecurity and Bioterrorism*, 1, 2003, 193–202.
82. K. Hoyt and S. Brooks, "A double-edged sword: globalization and biosecurity," *International Security*, 28(3), 2003–2004, 123–148.
83. A. Schuler, "Billions for biodefense: federal agency biodefense funding, FY2001-FY2005," *Biosecurity and Bioterrorism*, 2(2), 2004, 86–96, available online at: www.niaid.nih.gov/biodefense/ (accessed 21 September 2004).

Chapter 6

Stopping Nuclear Terrorism: The Dangerous Allure of a Perfect Defense

By Michael Levi

A successful nuclear attack by terrorists would be catastrophic. Intense fears of nuclear terrorism have led to a search for a perfect defense: destroying all terrorist groups that threaten the United States, sealing U.S. borders against loose nukes, or locking up all existing nuclear weapons and materials. Yet none of these strategies is a silver bullet. It is fantasy to believe that terrorism can be eliminated or that thousands of miles of U.S. borders—not to mention the borders of U.S. allies—can be sealed. Initiatives to secure nuclear weapons and materials are vital, but they will always fall short, too.

Rather than search for a perfect defense, which will never exist, counterterrorism strategists must use the many imperfect tools at their disposal to confront the many imperfect terrorist groups that they face. To pull off a nuclear attack, a group would need to acquire nuclear materials or a weapon, build a bomb or unlock an existing one, move that weapon to its target, and detonate it. Securing nuclear weapons and materials, although critical, confronts only one part of a plot and cannot eliminate the threat entirely. Strategists must build on this one defense to develop an integrated defensive system that also draws on border security, law enforcement, intelligence operations, military and diplomatic initiatives, and emergency response efforts. To do so properly, they must develop a more realistic picture of nuclear terrorism that draws on a careful understanding of how terrorist groups work and how their plots can fail.

Fighting a Mirage

When strategies for preventing nuclear terrorism rely on silver bullets, less dramatic—but nonetheless crucial—measures are neglected. The search for a perfect defense is partly driven by outsized fears of terrorists' capabilities and the assumption that a worst-case, or "perfect storm," scenario will occur. But terrorists do not have superhuman powers; their plots are imperfect and contingent and can be derailed. Consider the analogy of a police department seeking to prevent bank robberies. If the department assumes that all thieves have cars that travel 200 miles per hour, the department will give up on planning carefully for car chases and focus almost entirely on guarding the banks. If it instead realizes that many thieves will have cars that travel only 100 miles per hour, it will also carefully develop tactics for chasing down robbers. Realistically assessing the full spectrum of possible threats—in this case, from Ferraris to Ford Escorts—spurs broader and more careful planning by the police department. The same would be true of the U.S. government's homeland security and counterterrorism policies if Washington adopted a more nuanced view of the nuclear terrorist threat.

Moving away from worst-case assessments of the capabilities of nuclear terrorists will require strategists to rethink many basic assumptions. Terrorist groups are limited in their capabilities. Some terrorist groups, for example, lack expert personnel but have extensive resources. Analysts generally assume that wealthy groups will use their resources, whether money or connections, to recruit the right people. But the case of Aum Shinrikyo, an apocalyptic Japanese cult, reveals that this logic is not always correct. The cult was both wealthy and well connected in the Russian bioweapons world, but it decided that expanding its circle of experts would threaten its cover and failed to pull off a successful anthrax attack. Defenses that might work against groups with limited capabilities should not be dismissed, even if they are likely to fail against others.

Terrorist groups also make mistakes. In the typical portrait of nuclear terrorists, if they see an opportunity to succeed, they will manage to seize it. But things do not always go as planned. A group skilled enough to acquire a nuclear weapon will not, for example, necessarily succeed in the simpler task of hiding it from authorities. In June 2007, a group of highly educated Britons attempted to build and detonate two car bombs in London. They succeeded in building the bombs, but their education and intelligence did not stop them from leaving their cars in a no-parking zone, a seemingly avoidable mistake that led to their capture. It is impossible to predict where groups will err, but defenses should be devised so that they are ready to exploit the mistakes that terrorists will almost inevitably make.

Another common misperception is that apocalyptic terrorist groups care little about risk or failure. Flashes of creativity, such as al Qaeda's use of airliners as missiles, have convinced many observers that such groups are invariably nimble and innovative. But many ambitious and successful terrorist groups are conservative and averse to failure—and thus stick to bread-and-butter tactics whenever possible, even when a new approach seems promising. Although 9/11 has convinced many that al Qaeda is innovative, it also provides evidence for the opposite conclusion: the group used battle-tested techniques for financing and travel. Defensive measures that target the most obvious terrorist tactics should not be dismissed simply because analysts can dream up ways in which those measures might theoretically be evaded.

It would be wrong and dangerous to replace a hysterical overestimation of terrorists' capabilities with an excessively optimistic underestimation. But a realistic assessment of the full range of terrorist capabilities would encourage counterterrorism strategists to abandon demands for a panacea and instead adopt defenses that effectively targeted a wide spectrum of terrorists' weaknesses. Such an approach is sorely needed, because even the best hope for a silver bullet against nuclear terrorism—locking down all nuclear weapons and materials—will always fall short.

The Lockdown Fallacy

The argument for a defense based on nuclear lockdown is simple. Nuclear weapons require either highly enriched uranium or plutonium. These materials do not occur naturally, cannot be produced by terrorist groups, and are found at only a small number of sites worldwide. In theory, therefore, governments can prevent nuclear terrorism by locking up the world's dangerous nuclear supplies and stopping new states from acquiring their own. In reality, unfortunately, it is not so simple. Much of the world's nuclear material is used regularly in military and civilian activities, which means it cannot literally be locked up—there are many people authorized to handle and transfer it who might also have the opportunity to divert it for nefarious uses. It is also difficult to stop every state from acquiring nuclear weapons or to persuade states to disarm once they have joined the nuclear club. Many observers believe that Iran will eventually acquire a nuclear bomb, and there is no guarantee that North Korea will give up its entire arsenal. Nor is it safe to assume that these states can—or want to—keep their nuclear materials out of terrorists' hands. Efforts to lock down nuclear materials and stem proliferation will sharply lower these risks—and should be pursued more vigorously than they currently are—but they will never produce absolute security. It is thus crucial to develop a defense that also attacks the remaining dimensions of nuclear terrorism.

Approaching the problem from multiple angles would sharply reduce the likelihood of an attack. A terrorist group trying to pull of an attack might need to, among many other things, recruit trained scientists and engineers, acquire nuclear materials on the black market, purchase nonnuclear weapons components, and smuggle a bomb across borders. Each of these elements may seem manageable on its own. But taken together they can look much more daunting. Even a group that has an 80 percent chance of successfully carrying out each of these four steps has at best a 40 percent chance of completing the whole process and mounting a successful attack. Accordingly, chipping away at a group's odds of success at each stage of its plan—in this case by creating programs to ensnare rogue nuclear scientists, increasing the frequency of sting operations in the market for nuclear materials, monitoring terrorist finances to spot purchases of nonnuclear components, and installing radiation detectors at border crossings—would have a strong combined effect. If each of these changes shaved a mere 25 percent off the odds of success at each of those four stages of a terrorist plot, that would lower the chances of a successful plot to 10 percent—and perhaps convince some terrorists to abandon their ambitions.

Squeeze Play

But a defense conceived of as a sequence of hurdles—what analysts call a layered defense—is only the beginning. Each defensive element has two purposes: to defeat terrorist plots and to force terrorist groups to alter their plans in ways that introduce new vulnerabilities for them. Forcing terrorist groups to adapt in ways that expose new weaknesses is the key to an integrated approach to counterterrorism. Just as a triathlete who overexerts himself in the water ends up struggling during the subsequent bicycle race, a terrorist group that pours its resources into acquiring a nuclear weapon may have a tougher time moving that bomb to its target. Counterterrorism strategists must see themselves as the racecourse designers: if they want to slow down the racer during the bike leg, they can make the swim more difficult. By making the run harder, they can force the triathlete to take the bike leg slowly in anticipation. Likewise, if terrorist groups must make sacrifices in one part of their plot in order to enhance their chances of success in another, an approach that squeezes them from every side will substantially reduce their ability to evade defenses—or their willingness to even try.

Such an approach might have helped foil the 9/11 attack. By deciding to fly hijacked airliners into buildings rather than blow up airplanes—as it had plotted to do in an attack over the Pacific in the mid-1990s—al Qaeda bypassed the challenge of smuggling explosives past airport security. On the other hand, it had to send operatives to flight school, which increased its exposure. Bomb detectors at airports forced the 9/11 plotters to improvise and adopt tactics that differed sharply from those historically used by hijackers. If the bomb detectors had been complemented by more careful monitoring of flight schools (as an ignored FBI memo recommended), the plot could easily have unraveled.

An integrated defense would similarly squeeze nuclear terrorists. Consider a terrorist group based in the Middle East that is attempting to acquire fissionable material for a nuclear weapon. The group first considers stealing highly enriched uranium from a well-guarded facility in Russia by cooperating with disgruntled workers there. It judges that some workers will be susceptible to bribery but fears exposure due to the high level of security. Abandoning this plan, the group then considers attacking the facility directly. But the prospect of a rapid law enforcement response discourages it. As a last resort, the group looks to the black market, but there it faces the risk of a government-run sting operation. This forces the group to either improvise yet again, and expose itself to new risks, or abandon its plot altogether. In this particular case, strong security for nuclear materials would drive terrorists into the hands of black-market sting operations, and the threat of sting operations would send terrorists back to face guards at the nuclear facility's gates.

Securing the Borders?

This way of thinking—looking at counterterrorism as an integrated system contending with imperfect terrorists—is particularly illuminating when it comes to border security. Analysts and politicians frequently point out that nuclear terrorists are likely to seek weapons-grade uranium metal because it is relatively easy to use in bombs and shielding it with lead or other substances can conceal it from even the best radiation detectors that the United States could install at its ports—and hence conclude that efforts to improve port security will be ineffective. But this conclusion is not a sound foundation for defensive planning. Faced with stronger controls over nuclear materials, a terrorist group may decide to exploit any opportunity to acquire nuclear weapons or materials that comes its way—and that may involve material that requires heavier or more unusual shielding to hide its radiation or entire weapons that are more conspicuous simply by dint of their weight or size. If counterterrorism strategists focused equally on these lesser but still important threats, they could design cheaper and quicker approaches to scanning incoming cargo that would allow authorities to start effectively screening a much larger number of shipping containers arriving at U.S. ports. Port security may not be particularly useful on its own—and it is still likely be ineffective against many threats—but when properly focused and combined with materials security, it can reduce the likelihood of a terrorist attack.

Such an integrated strategy would also introduce new problems for terrorists trying to smuggle nuclear material across the United States' porous borders. The case of Gazi Ibrahim Abu Mezer is illuminating. Abu Mezer entered the United States along a remote stretch of the U.S.-Canadian border as part of a plot to bomb subway tunnels in Brooklyn. He tried to do this twice, but each time he was caught and sent back. He was stopped again on his third try, but that time the U.S. Border Patrol allowed him into the United States. (The same pattern of trial and error is common with illegal immigrants on the U.S.-Mexican border.) If border guards both recognized such patterns and were equipped with simple radiation detectors, a future Abu Mezer carrying nuclear material would instead be detained.

The potential for materials and weapons security to work in concert with border controls runs even deeper. Tripling the number of guards at a nuclear facility may not stop determined terrorists from getting nuclear materials, but it would sharply reduce their chances of escaping unnoticed—and if nuclear thefts were reported promptly, border security could be stepped up. Strategists who expect nuclear lockdown to do the job by itself will be caught unprepared if that strategy fails, but pursuing both approaches simultaneously could foil a plot. Indeed, border security—and many other tools—can be much more useful than is often assumed when it is combined with materials and weapons security as part of a broader integrated defense.

A New Strategy

An integrated approach to defending the United States against nuclear terrorism will require institutional and bureaucratic changes. As part of a move toward a more coherent and well-grounded homeland security and counterterrorism policy, U.S. officials must mandate a new strategic intelligence assessment of the nuclear threat. This assessment should combine expertise on nuclear weapons with expertise on terrorism, disciplines that, according to the final report of the president's Weapons of Mass Destruction Commission (released in 2005), have often failed to work together. Moreover, since so many ideas about nuclear terrorism are deeply ingrained, the effort must include analysts who have never studied nuclear terrorism in the past and can therefore bring fresh ideas to the table. Rather than focusing narrowly on the United States' vulnerabilities, this assessment should also catalogue opportunities for exploiting terrorists' weak points in order to derail and deter nuclear plots.

That intelligence assessment should form the foundation for ongoing "red team" exercises, in which the government tasks certain individuals with defeating and exposing holes in existing defenses. Red-teaming tests the ability of defensive planners to counter threats that they might otherwise ignore or fail to anticipate. Such exercises must not emphasize the most dramatic threats while ignoring lesser ones. When governments test defenses by trying to defeat them by any means possible, they implicitly emphasize worst-case scenarios. Although this is a necessary tool for auditing counter-terrorism efforts, it is misleading when used by itself. Government Accountability Office red-teaming tests of U.S. border security, for example, have found important holes in U.S. defenses. But they have also obscured the potential of many defensive strategies that could target lesser but still realistic threats. In so doing, they have convinced many members of Congress that those strategies are largely useless.

Active efforts to integrate defenses are also essential. But many of the tools needed for fighting nuclear terrorism are not nuclear-specific: instead, they span the full range of homeland security and counterterrorism. That makes it impossible to cordon of nuclear terrorism and appoint a single powerful official to fully orchestrate a strategy against it. Multiple attempts to do so have sparked turf wars and failed. An effort to create a national domestic nuclear defense office in the White House in 2004 ran into insurmountable bureaucratic resistance; the new National Counterproliferation Center, housed in the Office of the Director of National Intelligence, is barred from coordinating efforts to secure nuclear weapons and materials so as to prevent it from stepping on the Department of Energy's toes; and the National Counterterrorism Center, also in the ODNI and perhaps the likeliest candidate for lead coordinator, faces sharp limits on its power to mandate action.

Still, coordination on a narrower scale is essential and feasible. The recently established Domestic Nuclear Detection Office, located in the Department of Homeland Security, already plays a useful if limited role in coordinating nuclear detection efforts. The Global Initiative to Combat Nuclear Terrorism provides a forum for international discussion on aligning counterterrorism efforts. A lead senior official reporting directly to the president should be appointed to coordinate nuclear-specific elements of a defense. That official should also contribute to broader counterterrorism and homeland security planning. Meanwhile, the National Counterterrorism Center must create a strategic operational plan, as it did for combating terrorist travel. This would allow each branch of the government to evaluate its efforts not in isolation but in the context of a broader defense.

U.S. policymakers will also need to carefully reassess their approach to secrecy. For example, many of those on the frontlines of the war against terrorism are local law enforcement officers who do not have high-level security clearances. Joint counterterrorism task forces have been established, and their members have better access to information, but these groups are relatively small and typically operate in response to intelligence about a specific plot. Empowering beat cops to spot telltale signs of a nuclear plot in the course of their normal activities may require sharing more information about nuclear terrorism with them. There are, of course, dangers involved in sharing secrets, but a careful reassessment of secrecy policies is long overdue.

Predicting the threat of a nuclear terrorist attack is far from an exact science, and the consequences of writing off such a threat could be dire. But overestimating terrorist capabilities and dismissing certain defensive tools because they are imperfect is dangerous, too. The United States' strategy to combat nuclear terrorism must strike a balance between these two extremes. Staging a nuclear attack is harder than most people think, and strategists and policymakers must adopt a more realistic and nuanced assessment of the threat they face. Ratcheting down the threat assessment in this way should actually spur them to do more, not less. Terrorist groups can be vulnerable at any stage of their plots, and attacking each of these vulnerabilities presents an opportunity for counterterrorism officials to tighten the screws.

Chapter 7

Confronting Right-Wing Extremism in the USA: The Far Right and Terrorism

By George Michael

Terrorism and other forms of political and social violence have long been associated with the far right. During the mid-nineteenth century, members of the reactionary Know-Nothing movement were occasionally involved in violent confrontations with Catholics and immigrants, most notably in the riot in New Orleans, Louisiana in 1858.[1] Right-wing violence in America reached its high-water mark during the Reconstruction Era Ku Klux Klan's reign of terror in the aftermath of the Civil War. The total amount of carnage that the hooded order perpetrated during that period is difficult to quantify, but it is estimated that in the state of Louisiana alone it was responsible for at least 2,000 killed, wounded or injured in the few weeks preceding the presidential election of 1868. In addition other estimates of violence for that period include seventy-five killings reported in Georgia, 109 in Alabama, and more than 150 for one single county in Georgia.[2] The primary victims were Black freedmen, "scalawags," "carpetbaggers" and Radical Republicans. The Klan's membership is estimated to have reached 555,000 during the Reconstruction period.[3] The Second Era Ku Klux Klan, which reached its zenith in the 1920s, had more than its share of violent episodes as it targeted primarily Blacks, Catholics and "morally lapsed" White Anglo-Saxon Protestants.

According to one estimate there have been 245 incidents of right-wing terrorism since 1978.[4] Although other forms of organized violence have punctuated American history, such as episodes from the political left and organized labor, right-wing violence appears to have a longer history and a more enduring quality.

Patterns of Right-wing Terrorism

It is very difficult to quantify with precision the incidence of terrorism in the United States. The American criminal justice system treats terrorism as a crime. When caught, terrorists are tried in regular courts as there is no special classification for the crime of "terrorism." The FBI has endeavored to classify and collate terrorist incidents in America. However, much of its data on domestic terrorism are inconsistent and unreliable due to changing classifications of terrorist incidents.[5] What's more, as a 1976 government report demonstrated, the sheer number of violent incidents of a political or quasi-political nature make this effort all the more cumbersome.[6] Finally, due to the wide range in the severity of violent incidents—ranging from scuffles with police and minor injuries to bombings and homicides—it is difficult to ascertain and operationalize an overall level of seriousness or threat posed from a particular variant of political extremism in comparison to others.

According to the Israeli political scientist, Ehud Sprinzak, right-wing terrorism is characterized by a process of "split-deligitimation," in which not only the "outsider" (e.g. foreigners, ethnic and religious minorities, reds, and gays) is targeted but contemporaneously, the very state itself, which is seen as ineffective, or worse yet, under the actual sway of the outsiders. This theoretically leads to an evolution of right-wing terrorism. At first the terrorists usually avoid confrontations with authorities and their animus is directed at the "outsider." However, eventually they convince themselves that the government is not doing enough to protect the "original community" and the state also becomes a target.[7]

Significant Episodes of Right-wing Terrorism

The Greensboro Massacre

If one had to pinpoint a particular incident that really marked the genesis of the revolutionary orientation of the far right, it would be the Greensboro Massacre in 1979. On November 3 of that year, members of a neo-Nazi party, the National Socialist Party of America, and a local Ku Klux Klan organization clashed with demonstrators led by members of the Communist Workers Party in a "Death to the Klan" rally. The confrontation had been a culmination in a series of disputes between the two sides. At the demonstration a shootout between the two sides ensued, in which five members of the Communist Workers Party were fatally wounded. The Klansmen and Neo-Nazis suffered no serious casualties.

In two subsequent trials—one local and one federal—juries acquitted all of the defendants.[8] The juries were convinced by the self-defense arguments put forth by the defense counsels. This clash was significant because it was the first high-profile incident of right-wing violence since the Civil Rights era. Moreover, it demonstrated that Klansmen and neo-Nazis could cooperate. Previously, Klansmen looked askance at the foreign and un-American orientation of the neo-Nazis. Henceforth, neo-Nazism would have a significant influence on the ideology of a much broader portion of the racialist right. For many in the movement, Greensboro was seen as an event equal in symbolic significance to Lexington and Concord.This event would also be the catalyst for the creation of a progressive-oriented watchdog group, the National Anti-Klan Network, which would go on to become the Center for Democratic Renewal.

Gordon Kahl and the Posse Comitatus

The farm crisis of the mid-1980s provided a seedbed for right-wing extremism in America's heartland. One of the more prominent organizations in that area of the country was the Posse Comitatus. Despite its sometimes bombastic rhetoric, the Posse posed only a minor irritant to authorities. However, that changed on February 13, 1983, when a Posse affiliate and radical tax protestor, Gordon Kahl, got embroiled in a confrontation with federal authorities in Medina, North Dakota. A group of lawmen sought to serve Kahl a warrant for tax violations. Extremely distrustful of authorities, Kahl refused to be served and a shootout ensued in which two marshals died and four others were wounded, including Kahl's son Yorie. Amazingly, Kahl a sixty-three-year-old farmer and World War II veteran, had single-handedly caused the authorities to retreat.

Kahl evaded authorities for about four months, but on June 3, 1983, they finally caught up with him in Lawrence County, Arkansas. Still defiant, Kahl managed to mortally wound a local sheriff, who also happened to fire a fatal shot which struck Kahl in the head. Not realizing that Kahl was dead, the lawmen attempted to force Kahl outside of his bunker dwelling by pouring fuel down the chimney. The place went up in flames and sparked rumors that the FBI had summarily executed Kahl then incinerated his corpse and murdered the slain sheriff to cover up the truth.[9] As a result, Kahl entered the far right's pantheon of martyrs. His death also became the catalyst and a call to arms of an underground right-wing terrorist group, the Order.

The Order

Shortly after the death of Gordon Kahl, an annual Aryan Nations Congress was held at the compound in Hayden Lake, Idaho in the summer of 1983. At that meeting a young charismatic member of the National Alliance, Robert Jay Matthews, hatched the idea that an underground resistance group ought to be created to avenge the death of Kahl. Matthews had considerable powers of persuasion and was able to ultimately draw nearly fifty members into his clandestine terrorist group, "the Order."[10] It went on a crime spree in the early 1980s, which included several armored car heists, robberies, bombings, and at least five homicides.[11]

Matthews reportedly drew much inspiration for his organization from a novel, *The Turner Diaries,* written by his ideological mentor Dr. William Pierce (under the pseudonym Andrew Macdonald), the chairman of the National Alliance, who died in 2002. Reportedly Matthews made the novel required reading for all members. It is perhaps the most widely read book in the subterranean world of the far right, and has sold approximately 350,000 copies—an amazing figure for an underground book.[12]

Like the "Organization" in *The Turner Diaries,* Robert Jay Matthews endeavored to build a clandestine resistance group. His group went on a crime spree and a terrorist campaign that gained nationwide notoriety. Its activities included counterfeiting, armored car heists, bank robberies, and four homicides. In a milieu in which terrorists often resemble more the gang that couldn't shoot straight than professional terrorists, the exploits of the Order were, in a word, electrifying. One armored car heist took in a whopping $3.6 million—which at that time in 1984 set the record for the highest amount of money ever stolen in such a robbery.

Although, the Order was racist and anti-Semitic in orientation, consistent with Sprinzak's theory of split deligitimation, it gave its highest priority to targets such as the state and other institutions. Matthews instructed Order members to avoid petty conflicts with racial minorities, as that would distract the group from its primary mission. A list of prominent enemies marked for assassination was compiled, which included the leader of the Southern Poverty Law Center, Morris Dees, the former Secretary of State, Henry Kissinger, banker David Rockefeller, television producer Norman Lear, and international financier Baron Elie de Rothschild.[13] Despite such lofty intentions, the Order settled for a Denver-based Jewish disc jockey, Alan Berg, as its first target of assassination. Berg was an acerbic talk radio host and occasionally berated far-right callers on his radio program.

The Order's exploits soon caught the attention of authorities and the FBI identified the group as the most serious domestic terrorist threat in the country.[14] Ultimately, the counterfeiting operation led to the group's demise, as one of its recruits, though not an official member, Tom Martinez, agreed to become an informant for the FBI after his arrest for passing counterfeit money that the Order had printed. He set up two of his colleagues, including Matthews, in a sting operation at a hotel. A shootout ensued, and amazingly Matthews escaped after wounding an officer.

He remained undaunted and issued a "Declaration of War" against the United States government, which he sent to several newspapers. Finally, authorities caught up with Matthews at Whidbey Island in Washington state. Matthews, however, refused to be taken alive, and a standoff which lasted a couple of days followed. He single-handedly engaged in several shootouts with SWAT teams. Eventually, the authorities lost their patience and on December 8, 1984, dropped white phosphorous illumination flares onto the roof of the house in which Matthews had barricaded himself. This set off a fire that engulfed the structure and Matthews perished in dramatic fashion. A concerted effort by federal, state, and local law enforcement agencies eventually crushed the Order, and many of its members are now serving lengthy prison sentences.

The Order's campaign caught many in the radical right by surprise. Some criticized their exploits as ineffectual and quixotic, while others lionized them as exemplary "Aryan warriors."

Although the Order tactically did not really achieve much, it was significant insofar as it marked a change in the orientation of the far right. The United States government was now seen as the enemy and the far right began to take on a more revolutionary posture. No longer did it seek to preserve the status quo. Rather it sought the overthrow of the US government, which it reasoned was now under the heel of the Zionist Occupation Government (ZOG). The Order has been lionized by the far right and its incarcerated members are regarded as POWs to others in the movement.[15]

The Covenant, Sword, and the Arm of the Lord

In the late 1970s a Christian Identity minister, Jim Ellison, founded a community in Arkansas known as the Covenant, Sword and Arm of the Lord (CSA). Over the years it took on a paramilitary orientation, as members believed that eventually their enemies would one day besiege their compound, which in a sense they did. Ellison stockpiled many weapons and trained members how to use them in a mock village called "Silhouette City."[16] Furthermore, the compound came to be seen as a safe haven for those in the far-right underground who sought to evade authorities.

The CSA was linked to several episodes of terrorism. For example, Ellison and another member, Richard Wayne Snell, firebombed a Jewish community center in Bloomington, Indiana, and a gay church in Springfield, Missouri.[17] In another incident, Snell fatally shot a pawnbroker in Texarkana, Arkansas. Finally, during a routine traffic stop, Snell opened fire on an Arkansas state trooper and fatally wounded him. The CSA also planned several other serious terrorist attacks but failed to successfully carry them out.[18]

The violence and terrorism emanating from the CSA compound eventually captured the attention of federal authorities. Once authorities had implicated the CSA in these offenses, the FBI's elite anti-terrorist unit, the Hostage Rescue Team, quickly surrounded its compound. Although the CSA demonstrated a proclivity for violence, after some negotiation, Ellison surrendered to authorities without incident. He eventually cooperated with authorities and turned state's evidence against some members of the Order in a trial at Fort Smith, Arkansas in 1988.

The Rise of "Leaderless Resistance"

After the demise of the CSA and the Order, the revolutionary right went into a period of retrenchment and did some soul-searching on the topic of revolutionary strategy. From this interlude emerged a change in the tactics of right-wing terrorists. The lessons drawn were that when a terrorist organization grew to the size of the Order, it would eventually fall prey to infiltration and eventually be crushed. The CSA demonstrated that it was not feasible for terrorist groups to congregate in a compound that could easily be identified and surrounded by authorities.[19]

The violence-prone far right now employs an approach known as "leaderless resistance," a kind of lone wolf operation in which an individual, or a very small cohesive group, engages in acts of anti-statist violence independent of any official movement, leader or network of support.[20] Inasmuch as the contemporary far right is organizationally fragmented, leaderless resistance makes a virtue out of necessity. Moreover, this notion dovetails well with the new internet technology.[21] However, a word of caution is in order. Although right-wing advocates of the leaderless approach have theorized extensively on the concept, it remains in large part a construct of academic scholars and journalists. That is to say, it is a somewhat facile way of attributing the often unorganized and sporadic nature of right-wing violence to some larger operational plan. There is anecdotal evidence which suggests that several perpetrators who appear to have employed this approach, were psychopaths with little ideological sophistication. That said, the leaderless resistance concept should not be completely dismissed as merely a topic for abnormal psychology. As one observer noted: Leaderless resistance arises in large measure because of the failure of organized right-wing terrorism.

Although a sign of desperation, the notion of leaderless resistance should not be taken lightly.[22] According to the government's own prosecution case, the most lethal act of domestic terrorism—the Oklahoma City bombing—appears to fit in the leader-less resistance category.

The Oklahoma City Bombing

On April 19, 1995, the Murrah federal building in Oklahoma City was bombed, leaving at least 168 dead and many others wounded. This attack was the most lethal act of domestic terrorism ever perpetrated on American soil. The subsequent investigation would implicate Timothy McVeigh as the chief culprit of the attack. He also had some accomplices including Terry Nichols and Michael Fortier; however, the scope of their involvement is still uncertain. In the FBI's own description, "one of its most extensive investigations" failed to turn up a significant militia connection to the bombing. McVeigh and Nichols are reported to have attended meetings of the Michigan Militia, but the group did not welcome them. To the contrary, members of the group thought they were loose cannons and should not be granted membership.[23]

Although McVeigh may not have had any formal affiliation with extremist groups, there is evidence to suggest that he was a denizen of the subterranean world of the far right. McVeigh appears to have found several reasons to carry out his attack. During the 1993 siege of the Branch Davidians, McVeigh made a pilgrimage to Mount Carmel. He was reported to have been extremely upset over how the government handled the situation. He was also angered by the government's siege at Ruby Ridge, Idaho at which federal agents shot and killed Randy Weaver's wife Vicki and their son Samuel. McVeigh had chosen the Murrah federal building in particular, because he wanted to punish the Bureau of Alcohol, Tobacco, and Firearms (BATF) and the FBI, both of which were involved in the two incidents mentioned above.[24]

In his personal letters and statements, McVeigh echoed many ofthe Constitutionalist themes that are popular in the Christian Patriot/Militia movement. He absorbed much of the anti-government propaganda of the Patriot movement at various gun shows that he often attended. Although he did not usually evince a visceral racism, he actively explored the propaganda of the racialist right. At one time, he obtained a trial membership of a Ku Klux Klan organization based in North Carolina. However, according to McVeigh he declined to renew his membership because he felt that the Klan was "manipulative to young people." Moreover, he felt that the government, and not racial minorities, was his enemy.[25]

The Aryan Republican Army

Around the same time of the Oklahoma City bombing, a six-man group of bandits, which called itself the "Aryan Republican Army," was making headlines in the Midwest for a series of bank robberies that confounded authorities. All totaled, the Aryan Republican Army is reported to have been responsible for robbing twenty-two banks and netting some $250,000 in cash. Although the goals of these renegades are still murky, they are alleged to have consorted with some of the more notorious figures in the far-right underground. One author goes so far as to theorize that the money stolen was used to fund the terrorist attack on the Murrah federal building in Oklahoma City.[26]

Authorities eventually closed in on the Aryan Republican Army and put an end to its campaign. Although its significance is disputable, it could presage a pattern in other countries beset by terrorism, where criminal enterprises use their ill-gotten gains to fund terrorist and revolutionary movements.

Lone Wolves

The far right is a relatively small and heavily stigmatized movement that does not enjoy broad-based support from the public. Most in the movement realize that the forces arrayed against them—the government and the watchdog organizations—are collectively vastly more powerful than they are. Consequently, there has always been a conservative majority in the far right that believed that it would be foolhardy to prematurely engage in revolutionary violence. Such an approach would almost certainly lead to organizational suicide. Thus the more conservative elements advocated a strategy that would concentrate on utilizing propaganda to build a revolutionary majority. This came to be known as the theory of mass action.[27]

A leading proponent of this approach was George Lincoln Rockwell, the founder of the American Nazi Party that was active in the 1960s. Rockwell believed that events such as racial integration, school busing, the Vietnam War, race riots and rising crime would engender urban mayhem and thus create favorable conditions for his party. He entertained the idea that his party could actually win national power by 1972. However, Rockwell fell to an assassin's bullet in 1967, and with his departure some elements of the far right became disillusioned with the conservative approach.

Joseph Tomassi, a member of Rockwell's successor organization, eventually departed and founded the National Socialist Liberation Front (NSLF), a neo-Nazi organization that patterned itself on the left-wing models of the Weathermen and the Symbionese Liberation Army. Tomassi correctly saw that in the early 1970s, the idea of creating a Nazi-style party that would win support of a majority of the population was futile. However, he believed that it was still possible to strike blows against "the system," provided that revolutionaries were prepared to act resolutely and alone. Whereas the state demonstrated over and over again that it could infiltrate and effectively neutralize any right-wing organization, it had yet to develop the capability of thwarting the actions of individuals or small groups acting alone. However, the NSLF campaign was reckless and its revolutionary arm was quickly crushed. Like Rockwell, Tomassi was also killed by a disgruntled member. Although the organization never succeeded in striking a serious blow against "the system," according to one observer its "contribution to the leaderless resistance concept [was] incalculable."[28] Thus was born the concept of leaderless resistance in the far right. However, the approach had still not been given a name and the idea would languish for the most part among the far-right theorists until the early 1990s.

Another resistance approach occasionally used by the far right has been the cellular model. The far right's version of this approach can be traced back to an anonymous tract titled The John Franklin

Letters, which was popularized by the John Birch Society in 1959.[29] The John Birch Society promoted the book as a call to resistance in the wake of a communist takeover, which the group believed was always just around the corner. Although the Birch Society organizationally patterned itself on the cellular model, it was loath to engage in any political violence.

The Minutemen also adopted the cellular structure with a centralized command structure, and some of its members were implicated in violence. Its leader, Robert DePugh, published a manual for activists, Blueprint for Victory, which gave advice on political, military, economic and psychological warfare.[30] The Minutemen tried to maintain the semblance of a centralized command structure, and discouraged individual initiative, leaderless-resistance types of operation.

Despite these efforts, some members of the Minutemen ran afoul of the authorities and the organization was effectively shut down after some episodes of violence. The popularity of the cellular model reached its zenith with the campaign of the Order. The Order carried out various missions in small groups, and there were plans on splitting the organization into interlocking separate cells. Despite some spectacular moments, it was resoundingly quashed after a concerted effort by government authorities. The crushing defeat of the Order and the CSA ushered in a debate on the best strategy of "resistance" to employ in the far-right movement. Thus the concept of leaderless resistance was reexamined and updated for current conditions.

Jeffrey Kaplan observes that the movement's discourse became increasingly chiliastic by the late 1980s.[31] The mass action theories of previous generations were seen as unrealistic. The cellular model had been discredited with the demise of the Order. A period of despair and hopelessness seemed to settle over the movement. However, two events, first the ambush of Randy Weaver's home in Ruby Ridge in 1992, and second, the siege at Waco in 1993, once again galvanized a broad segment of the far right. Previously isolated voices calling for resistance found their messages being heard by a more receptive audience.

Leaderless resistance had several proponents during the 1980s, but the concept was truly crystallized and gained currency as a result of the October 1992 meeting in Estes Park, Colorado convoked by Christian Identity minister Pastor Pete Peters. This event provided a forum for the articulation of a new leaderless resistance approach. Whereas prior to the meeting the concept was only vaguely recognized by some, it was now given a name and disseminated to a much larger audience. This event, more than any other, popularized the notion in the far-right community.

Conclusion

The modus operandi of right-wing terrorists has undergone some change over the past decade or so. The primacy of leaderless resistance in the racialist right presents a challenge for authorities insofar as it is difficult to predict from where a threat may come. It is much easier to infiltrate an organization rather than a clique of a few individuals, or to read the mind of one lone wolf. It is important to keep in mind that the prominence of leader-less resistance is symptomatic of the organizational and financial weakness of right-wing terrorists. No real terrorist infrastructure as such exists in the far-right milieu.

What is more, as James Adams has pointed out, the sine qua non for the development of such an infrastructure is money.[32] Generally speaking, right-wing terrorists have neither garnered support from state sponsors nor developed adequate means of self-financing.

In recent years organized groups in the Militia movement (e.g. the Montana Freemen and the Republic of Texas) have been a problem for authorities. However, the outlaws in this segment appear to concentrate their illegal activities primarily on "paper terrorism" (e.g. issuing phony checks and filing illegal liens) rather than political violence.

One obvious concern is the potentially greater lethality of instruments of terrorism. Changes in technology have enabled terrorists to inflict potentially greater damage. The Oklahoma City bombing demonstrated the potential lethality of "terrorism on the cheap" conducted by just a few individuals. Indeed, more so than the terrorists of the previous decades, contemporary terrorists appear more intent on punishing their targets.

Elements of the American far right have demonstrated on at least one previous occasion that they would not be averse to such tactics. For example, the CSA allegedly planned to poison water supplies, and recently a militia group, Southeastern States Alliance, allegedly planned on destroying a nuclear power plant. Fortunately, these examples never made it beyond the planning stages.

Another ominous trend is the targeting of vital parts of infrastructure that could potentially paralyze or do great damage to the day-to-day lives of many people. Finally, bio-terrorism, such as an anthrax attack, is one more measure that terrorists could employ.

Discussion Questions

1. What are some of the resistance movements that inspired Timothy McVeigh's bombing of the Oklahoma Federal Building?
2. Discuss the reasons why the radical right uses the approach of leaderless resistance and some of the failed approaches that preceded it.
3. Discuss some methods law enforcement might use when dealing with far-right terrorist groups and the new threats they pose.

Notes

1. See Beals, Carleton, *Brass-Knuckle Crusade: The Great Know-Nothing Conspiracy: 1820–1860* (New York: Hastings House Publishers, 1960) pp. 193 207.
2. Forster, Arnold and Benjamin R. Epstein, *Report on the Ku Klux Klan* (New York: Anti-Defamation League, 1965) p. 14.
3. Forster and Epstein, *Report on the Ku Klux Klan,* p. 14.
4. Hewitt, Chris, "Responding to Terrorism," unpublished manuscript, 2000, p. 5. According to Hewitt shootings accounted for 32 percent of the incidents, followed by bombings (28 percent), assaults (22 per cent) and other (18 per cent).

5. For a critical examination of the FBI's data collection on domestic terrorism see Hamm, Mark S., *Terrorism, Hate Crime, and Anti-Government Violence: A Preliminary Review of the Research* (Indiana: Indiana State University, 1996).
6. In 1976 the National Advisory Committee on Criminal Justice Standards and Goals issued a Report on the Task Force on Disorders and Terrorism, which listed violent incidents of a political or quasi-political nature in the period from January 1965 to March 1976. As the report demonstrates, violent incidents were practically daily occurrences throughout much of that period. Most of the incidents appear to emanate from the political left. See National Advisory Committee on Criminal Justice Standards and Goals, *Report on the Task Force on Disorders and Terrorism* (Washington DC: National Advisory Committee on Criminal Justice Standards and Goals, 1976).
7. Sprinzak, Ehud, "The Process of Delegitimation: Towards a Linkage Theory of Political Terrorism," *Terrorism and Political Violence*, 3 (1), pp. 50–68 (1991).
8. In a civil trial however, a jury awarded approximately $400,000 for the death of one of the five slain and the injuries of two of the wounded. They found five Klansmen and Nazis, a police informant, and two police officers liable. Wheaton, *Code Name Greenkil.*
9. Coulson, Danny O. and Elaine Shannon, *No Heroes: Inside the FBI's Secret Counter-Terror Force* (New York: Pocket Books, 1999) pp. 192–3.
10. The organization used several names including "the Silent Brotherhood" and a German version of that same title, The Brüder Schweigen.
11. Aho, James, *The Politics of Righteousness: Idaho Christian Patriotism* (Seattle: University of Washington Press, 1990) p. 7.
12. Segal, David, "The Pied Piper of Racism," *Washington Post*, January 12, 2000, C1, p. 8.
13. Coulson and Elaine Shannon, *No Heroes*, p. 194.
14. This is according to the statements of Danny O. Coulson, the founder of the FBI's Hostage Rescue Team, who was involved in the Order investigation. Coulson and Shannon, *No Heroes*, p. 195. According to one estimate, the investigation is said to have involved one quarter of the total manpower resources of the FBI. Aho, *The Politics of Righteousness*, p. 61.
15. The prison addresses of the "POWs" are occasionally listed in the far-right literature, and readers are encouraged to write and provide material and moral support to them and their families.
16. Coulson and Shannon, *No Heroes*, p. 212.
17. Coulson and Shannon, *No Heroes*, p. 222.
18. Some of the alleged plots included an attempt to blow up a natural gas pipeline in Fulton, Arkansas; however, the explosive failed. Some CSA members allegedly planned to assassinate several public offices, including those of federal judges H. Franklin Waters and Jack Knox, and also US Attorney Asa Hutchinson. Ellison and Noble planned to bomb a gay church in Kansas City, Missouri, but Noble lost his nerve and failed to carry out the attack. Finally, according to Ellison and Noble, the CSA planned to use a 30 gallon drum of cyanide that it had stockpiled to poison the water supplies of New York City and Washington DC. It was hoped that this would foment urban unrest and precipitate their anticipated revolution.
19. It took the far right quite a while to learn this lesson of how vulnerable they would be if isolated in a rural setting. As Smith and Damphousse observed, the far left had widely publicized in their publications the failure of Che Guevara's similar strategy in Bolivia. Smith, Brent L. and Kelly R. Damphousse, "Two Decades of Terror," in Kushner, Harvey W. (ed.) *The Future of Terrorism: Violence in the New Millennium* (Thousand Oaks CA: Sage, 1998) p. 142.
20. As described in Kaplan, Jeffrey, "Leaderless Resistance," *Terrorism and Political Violence*, 9 (3), p. 80 (1997).
21. As an example, a website operated by an anti-abortion activist in Oregon listed the names and addresses of doctors who performed abortions. The site contained unsubtle suggestions that there should be some kind of retribution against them.

See "Anti-abortion Web Site Goes on Trial," *USA Today*, January 7, 1999. http:// usatoday.com/news/ndswed05.htm. This is just one example of how this notion of leaderless resistance can work. Activists with no formal organizational ties and who don't even know one another can post information on the web, and others acting independently can take these cues and commit acts of terrorism as they see fit.

22. According to James Aho, roughly one half of all right-wing homicides from the period from 1980 through 1990 can be attributed to individuals acting alone. Aho, *The Politics of Righteousness*, p. 62.
23. Stickney, Brandon M., *All-American Monster: The Unauthorized Biography of Timothy McVeigh* (Amherst, NY: Prometheus Books, 1996) pp. 97–100, 158–9. One Michigan Militia member who was present at the meeting told me that McVeigh and Nichols were not well received because of their "extreme racial feelings" and were asked to leave. Interview with Norm Olson, September 8, 2000.
24. Michel, Lou and Dan Herbeck, *American Terrorist: Timothy McVeigh and the Oklahoma City Bombing* (New York: Regan Books, 2001) pp. 167–8.
25. McVeigh claims to have joined the Klan because he thought that it was "fighting for the restoration of individual rights, especially gun rights." After more research he discovered that the Klan was "almost entirely devoted to the cause of racism." Michel and Herbeck, *American Terrorist*, pp. 88–9. His naiveté of the true nature of the Klan seems a bit disingenuous insofar as he was of above average intelligence and intellectually curious. However, it is also worth mentioning that he appears to have been a "seeker" in the truest sense of the word and was thus willing to look practically anywhere for knowledge.
26. The criminologist Mark Hamm posits the theory that the Aryan Republican Army was part of a larger conspiracy and revolutionary division of labor, in which the bandits would use their money to fund right-wing revolutionaries. He strongly suspects that members of the Aryan Republican Army were instrumental in the Oklahoma City bombing. For more on his theory see Hamm, Mark S., *In Bad Company: America's Terrorist Underground* (Boston MA: Northeastern University Press, 2001).
27. I borrow much of the analysis in this section from Kaplan's article "Leaderless Resistance," *Terrorism and Political Violence*, 9 (3), pp. 80–95 (1997).
28. Kaplan "Leaderless Resistance," pp. 81–2.
29. *The John Franklin Letters* (New York: The Bookmailer, 1959). Robert Griffin, the biographer of Dr William Pierce, believes that Revilo P. Oliver was the book's author. See Griffin, *The Fame of a Dead Man's Deeds*, p. 143.
30. DePugh, Robert B., *Blueprint for Victory*, 4th edn (Norborne MO: Salon Publishing Company, 1978).
31. By "chiliastic" Kaplan refers to several themes, including a period of tribulation in which right-wing terrorists believe that they must engage in a violent struggle before they reach victory, which will result in a period of millennial bliss. Also, there is a Manichean theme to this struggle as the right wing sees themselves as a righteous remnant while their enemies are seen as the embodiment of evil. Kaplan argues that as the far right was banished from beyond the pale of respectability it abandoned any "reformist" ambitions that it once had:
32. The state ZOG was increasingly seen as not worth claiming, and with this conclusion, the movement's dreams became increasingly chiliastic. With this too, the pattern of violence emanating from the fringes of the movement began to shift from vigilantism to anti-state terrorism. (Kaplan 1995, pp. 85–7)
33. Adams, James, *The Financing of Terrorism* (New York: Simon and Schuster, 1986) in White, Jonathan R., *Terrorism: an introduction* (Stamford CT: Wadsworth, 2002) pp. 40–42.

References

"Anti-abortion Web Site Goes on Trial," *USA Today*, January 7, 1999, http://usatoday. com/news/ndswed05.htm.

Adams, James, *The Financing of Terrorism* (New York: Simon and Schuster, 1986) in White, Jonathan R., *Terrorism: an introduction* (Stamford CT: Wadsworth, 2002) pp. 40–2.

Aho, James, *The Politics of Righteousness: Idaho Christian Patriotism* (Seattle WA: University of Washington Press, 1990).

Beals, Carleton, *Brass-Knuckle Crusade: The Great Know-Nothing Conspiracy: 18201860* (New York: Hastings House Publishers, 1960).

Coulson, Danny O. and Elaine Shannon, *No Heroes: Inside the FBI's Secret Counter-Terror Force* (New York: Pocket Books, 1999).

DePugh, Robert B., *Blueprint for Victory*, 4th edn (Norborne MO: Salon Publishing Company, 1978).

Forster, Arnold and Benjamin R. Epstein, *Report on the Ku Klux Klan* (New York: Anti-Defamation League, 1965).

Griffin, Robert S., *The Fame of a Dead Man,s Deeds: An Up-Close Portrait of White Nationalist William Pierce* (self-published e-book, 2000).

Hamm, Mark S., *In Bad Company: America's Terrorist Underground* (Boston MA: Northeastern University Press, 2001).

Hamm, Mark S., *Terrorism, Hate Crime, and Anti-Government Violence: A Preliminary Review of the Research* (Indiana: Indiana State University, 1996).

Hewitt, Chris, "Responding to Terrorism," unpublished manuscript, 2000.

Interview with Norm Olson, September 8, 2000.

Kaplan, Jeffrey, "Leaderless Resistance," *Terrorism and Political Violence*, 9 (3), p. 80 (1997).

Michel, Lou and Dan Herbeck, *American Terrorist: Timothy McVeigh and the Oklahoma City Bombing* (New York: Regan Books, 2001).

National Advisory Committee on Criminal Justice Standards and Goals, *Report on the Task Force on Disorders and Terrorism* (Washington DC: National Advisory Committee on Criminal Justice Standards and Goals, 1976).

Segal, David, "The Pied Piper of Racism," *Washington Post*, January 12, 2000, C1, p. 8.

Smith, Brent L. and Kelly R. Damphousse, "Two Decades of Terror," in Kushner, Harvey W. (ed.) *The Future of Terrorism: Violence in the New Millennium* (Thousand Oaks CA: Sage, 1998).

Sprinzak, Ehud, "The Process of Delegitimation: Towards a Linkage Theory of Political Terrorism," *Terrorism and Political Violence*, 3 (1), pp. 50–68 (1991).

Stickney, Brandon M., *All-American Monster: The Unauthorized Biography of Timothy McVeigh* (Amherst, NY: Prometheus Books, 1996).

Wheaton, Elizabeth, *Code Name Greenkil: The 1979 Greensboro Killings* (Athens GA: University of Georgia Press, 1987).

White, Jonathan R., *Terrorism: an introduction* (Stamford CT: Wadsworth, 2002).

Chapter 8

The Blame Game on the Border: Perceptions of Environmental Degradation on the United States-Mexico Border

By Lisa M. Meierotto

Introduction

In June of 2008, I joined a Border Patrol agent stationed in Nogales for a "ride-along." We spent the first part of the day touring the detention facilities in Nogales, and I was able to view the massive amounts of guns and drugs regularly confiscated by Border Patrol. Countless men and women were held in detention, standing in long lines waiting to be fingerprinted, photographed, and eventually placed on the "Wackenhut" bus and sent back to the border.[1] The second half of the day was spent chasing Mexicans. We drove through the streets in and around Nogales, the agent valiantly searching for an "illegal" so that I would have an interesting story to write about in my dissertation. The roads outside Nogales and near the wall are full of potholes, and narrow roads wind up and down steep hills. We were racing around corners and speeding down hills, hoping to spot someone. He instructed me what to do if he had to leave the vehicle, how to cover my head if we were attacked by rocks, and how to work the radio in case something happened to him. It was simultaneously exciting and terrifying. After a couple of hours of this cat-and-mouse routine, we went back to the station. He was clearly disappointed that we didn't "catch" any Mexicans.

At the end of the day, we spent an hour or so watching the deportation process from atop a hillside in Nogales. After the apprehended individuals are loaded onto the bus, they are transported a short distance to a border check-point, and then released. They walk single-file back across the border, where food and soda stands dot the landscape, and eager *coyotes* await the arrival of their return customers.

Lisa M. Meierotto, "The Blame Game on the Border: Perceptions of Environmental Degradation on the United States-Mexico Border," *Human Organization*, vol. 71, no. 1, pp. 11–21.

Most will attempt to cross again, and most will eventually succeed. Some of these individuals will ultimately cross the border through environmentally protected areas in the United States, and as such unwittingly become a part of an entrenched environmental conflict—the degradation of wilderness-protected areas along the United States-Mexico border. While stories about undocumented border crossing often make the evening news and are subject to intense academic study, the environmental impact of border crossing has not yet been adequately studied.

The aims of this paper are two-fold: first, I endeavor to expand the study of the causes of environmental degradation in protected areas on the Arizona-Sonora international border. My goal here is to move beyond typical explanations of border degradation that tend to place blame on undocumented border crossers by also considering the environmental footprint of Homeland Security—in particular the apprehension activities of the Border Patrol. The second aim of the paper is to compare and contrast varied perceptions of environmental degradation in the borderlands. Thus, in the second half of the paper, I explore ways in which perceptions of environmental degradation are at times shaped and influenced by wider social and cultural beliefs. I argue that negative environmental perceptions of migration are both caused by and contribute to beliefs about people of Mexican-origin as dirty, threatening, and invasive. Through an analysis of the causes and perceptions of environmental degradation currently manifesting on the United States-Mexico border, we can find insight into the politics and social tensions present in immigration debates and conflicts. The concept of a "game" on the border has been employed before (e.g., Andreas 2009), and the cat-and-mouse game of pursuit along the United States-Mexico border is well known. To expand upon the game metaphor, the blame game in this paper refers to the shifting perceptions and accusations of *environmental* damage occurring as a result of immigration policies and enforcement practices.

Shifting Immigration/Homeland Security

Changes to immigration policy reaching back to the early 1990s have resulted in increasing levels of undocumented border-crossing and smuggling across Arizona's desert wilderness (Cornelius 2001; Martinez 2006; Nevins 2002; Regan 2010). The 1990s border policy changes effectively sealed borders in urban areas in California and Texas (Cornelius 2001). This created a funneling effect—as urban crossing zones were closed, immigrants and smugglers were forced to develop new northward routes, and the desolate and barren deserts of the southwest provided vast, open expanses of land to cross. While the funneling effect is well known among migration scholars and immigration policymakers, less attention has been paid to the impact of the corresponding border security buildup (see Urban 2008 for a relevant study on immigration, population, and security). Importantly, as more immigrants and smugglers cross through the fragile desert, there is a corresponding increase in the number of Border Patrol agents in the region. The security influx was further accelerated by the creation of Homeland Security and related policies after September 11, 2001 (Chavez 2008; Urban 2008). Each policy evolution has resulted in increasing human impact on fragile desert ecosystems. Unfortunately, the environmental impact of immigration policy changes and the subsequent buildup of Homeland security have not been sufficiently studied in the academic community. Some research has considered the environmental impact of population growth (e.g. Kraley 1995), but little attention has been paid to the direct environmental impact of undocumented immigration and subsequent security build-up.

The observations presented in this paper are based on research conducted primarily at Cabeza Prieta National Wildlife Refuge and in the adjacent town of Ajo, Arizona, between January 2007 and May 2009. The primary methods used were participant observation, informal interviews, and archival review. Secondary review of scholarly research and media sources were also utilized. Research informants included staff at Cabeza Prieta National Wildlife Refuge, staff members of other nearby land-use agencies, residents and visitors in nearby Ajo, Arizona, and with regional Border Patrol agents. Finally, I used remote sensing (aerial photography) and geographic information system (GIS) software to locate and document illegal roads.

Throughout the study period, care was taken to ensure safety of all involved. I made the decision that direct interviews with undocumented border crossers would not be conducted in an effort to preserve their anonymity and safety. In an ever-escalating context of crime rings and increasing violence, the researcher's safety was also taken under consideration, and, thus, data presented in this paper on migrants is gathered from secondary sources. There are inherent security risks faced by any anthropologist working in any border region (Donnan and Hastings 2010), but these risks are often amplified for female anthropologists in the United States-Mexico border region (Gauthier 2010).

Environmental Degradation at Cabeza Prieta National Wildlife Refuge: Joint Impacts of Migration and Security

In the state of Arizona, the international border transects the Sonoran Desert. The landscape is comprised of fragile ecosystems, biodiversity hotspots, and recreational magnets. It is a beautiful and intriguing landscape, home to many threatened and endangered species. National Monuments and Parks with Saguaro cactus forests attract thousands of tourists each year. Two large wildlife refuges[2] run along the border, the largest of which, Cabeza Prieta National Wildlife Refuge (CPNWR), shares 56 contiguous miles with the Mexican border. Cabeza Prieta is one of 300 Wildlife Refuges nationwide and is a nationally-designated Wilderness area. Cabeza Prieta means "Dark Head" in Spanish and refers to a dark-colored mountain on the western side of the refuge. The refuge was established in 1939 for the protection and management of desert resources. At 860,010 acres, it is the third largest wildlife refuge in the continental United States (outsized by the Desert Wildlife Refuge in Nevada and the Charles M. Russell Refuge in Montana). Adjacent areas to the refuge include a national park and the Tohono O'odham Nation to the east, the massive Goldwater Air Force range to the west, and Mexico to the south. On the Mexican side of the border, conservation programs are also in effect, but similar to the United States side, protected areas are fragmented and under decentralized management. Highway Number 2 in Mexico cuts the landscape from East to West across the Sonoran Desert and roughly parallels the international border.

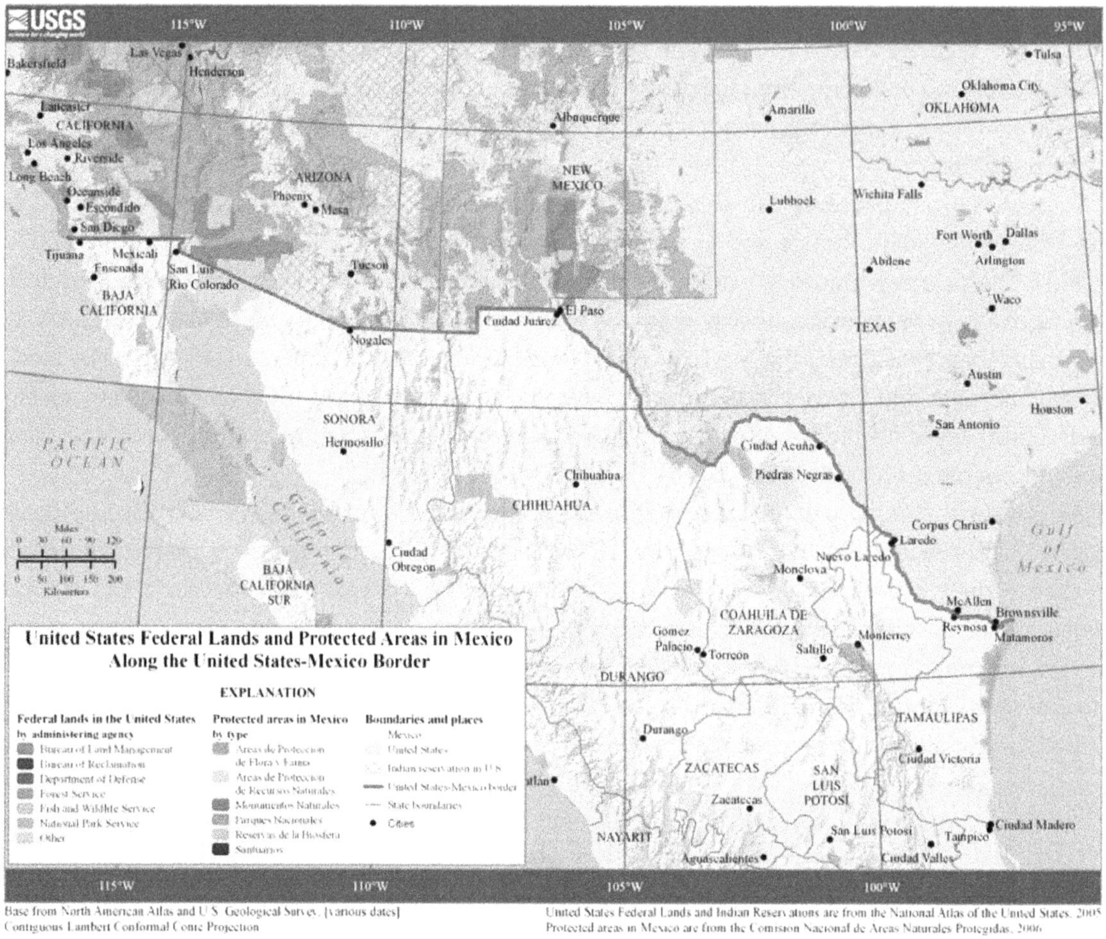

Figure 8.1 Map of protected Areas on the United State–Mexico Border

Source: U.S. Geological Survey. Copyright in the Public Domain.

As a result of immigration policy changes in the 1990s, Cabeza Prieta was thrust to the forefront of immigration problems, and the refuge became a hotbed of political, police, and military activity as northward-bound migrants began to cross through this desolate region, and smugglers found a wide-open range through which to transport their goods. Estimates of the number of migrants crossing through the region vary widely, with a high estimate claiming that in the early 2000s, 4,000–6,000 crossings occurred each month between the months of April and June (Slown 2003). In response, there has been a monumental influx of Border Patrol agents and state-of-the-art security technology, including numerous cameras and watchtowers and literal assault upon the landscape from the sheer amount of vehicular traffic, especially off-road in wilderness areas.The combined forces of immigration and border security (within the context of climate change bringing about a drier, hotter regional ecosystem) have the left the refuge in dire straights.

Cabeza Prieta National Wildlife Refuge is one of the most degraded wilderness areas in the United States (Lusk 2008; Public Employees for Environmental Responsibility 2008; Wilderness Society 2008). A wilderness area under United States federal designation should be devoid of most human impact, and especially untouched by motorized vehicles. But on a typical day out on the refuge, you would encounter a large number of vehicles. On the west side, you might see groups of people heading out to cruise around in their four wheelers or dune buggies. In addition to recreational visitors like hikers and bikers, there is another group of "visitors" passing through Cabeza Prieta. Each year hundreds and likely thousands of people cross through this refuge from Mexico en route to jobs and markets in the United States. Immigration through this area is brutal and often fatal. The Camino del Diablo has been described as the "deadliest immigrant trail in North America" (Annerino 1999: 9). While authors like Annerino (1999), Urrea (2004), Regan (2010) and Ferguson, Price, and Parks (2010) have tried to humanize the experiences of migrants in the region, the brutality and challenge of a trek through the Arizona desert is difficult to describe in text and not widely understood by people outside of the immediate locale. Because so many crossings and deaths are in remote locations, accurate statistics will likely never be available. One estimate from No More Deaths (Regan 2010) is that more than 5,000 migrant bodies have been found in the southwest between 1994 and 2009. As a result of immigration changes in the 1990s, a number of apprehensions have dropped (this is cited as a sign of success according to Border Patrol), but an individual's chance of dying became much higher in that same time period. By one estimate, a migrant is 17 times more likely to die crossing in 2009 than in 1998 (McCombs 2009). In the general border region, the number of migrant deaths has doubled between the mid 1990s and 2005 (GAO 2006:3).

But migrants are just a small portion of the thousands of people crossing southwestern deserts each year. In pursuit of migrants and smugglers, an equally significant force is sweeping across the desert—Border Patrol, law enforcement agencies, and military organizations, all garnered under the umbrella of Homeland Security. At the local level, there is virtually no dispute as to whether or not environmental degradation is occurring across Cabeza Prieta National Wildlife Refuge. Refuge staff. Border Patrol agents, and recreational visitors all attest to the fact that increases in traffic (both vehicular and foot) in recent years are undeniable and are creating long-term environmental damage. The primary underlying drivers of this degradation, however, are subject to great controversy. Blame tends to be assigned in one of two directions: either the migrants and smugglers are destroying the environment, or conversely, Border Patrol is to blame for the worst assaults on nature. Accusations abound, with fingers pointing at different user groups who access the refuge. Sorting these divergent accusations out requires analysis on two levels: first, determining what environmental damage is actually occurring, and second, after assessing environmental damage, discerning which user group holds the greatest responsibility for that damage.

While it seems logical that both increases (in number of undocumented border crossers and Border Patrol agents) would result in environmental degradation through increases in foot and vehicle traffic in wilderness areas, there is often great attention paid to the environmental impact of migrants—and this is evident in both mainstream media and academic literature. However, based on my own observations during fieldwork, I found that the environmental impact of Border Patrol was at least as significant, if not greater than the impact of border crossers. The following section describes the environmental impact of migration and Border Patrol.

Environmental Impacts of Migration

Large-scale migration is one of the most pressing threats to biodiversity conservation efforts today (Brechin 2003; Homer-Dixon 1999; Pirages and DeGeest 2004). It is estimated that up to 300 million people migrate across borders each year on a global scale (Oglethorpe et al. 2007). Environmental degradation from migration can occur at the origin site, through the loss of local place-based knowledge, at the destination site, through increases in population, and also along the migration corridor itself (Oglethorpe et al. 2007). Migration most often negatively affects biodiversity by contributing to a loss of species and genetic diversity, habitat loss and fragmentation, pollution, and the spread of invasive species (Oglethorpe et al. 2007). The International Symposium on the Environmental Impacts Resulting from Mass Migrations found four general categories of negative environmental impacts from migration (adapted from Mc-Intyre and Weeks 2002), naming those that:

- directly damage ecosystems (i.e., roads)
- indirectly affect local markets and prices (i.e., enforcement costs)
- indirectly affect environmental heath conditions (i.e., human waste)
- cause political impacts (i.e., tensions between local residents and migrants)

Conservationists have not thoroughly considered the impacts of migration on conservation efforts and policies (Oglethorpe et. al. 2007).

While conservation scientists may not have adequately considered migration impacts, media outlets and policymakers have drawn great attention to the issue of migration in southern Arizona (e.g., Clarke 2008). One popular perception in the mainstream media explored in this paper is that migrants and smugglers leave a devastating environmental footprint on our nation's protected pristine areas (e.g., Fox News 2008; Nealson 2009). Migrants are seen as "dirty," leaving behind food and human waste and the ubiquitous discarded plastic water jug. This "trash" viewpoint is also at times propagated by the United States government. Former United States Secretary of Homeland Security, Michael Chertoff summarizes this perception in a media interview: "Illegal migrants really degrade the environment. I've seen pictures of human waste, garbage, discarded bottles, and other human artifacts in pristine areas. And believe me, that is the worst thing you can do to the environment" (Associated Press, October 1,2007).

Migrants are often identified as causing environmental degradation on the border, in both the mainstream media and among local residents. For example, one recreational visitor asserted that the refuge is "not as nice as it used to be prior to all the illegals." However, the actual impact of garbage (primarily plastic bottles, discarded clothing, backpacks, and food) on border-area conservation sites has not been documented empirically—in other words, no calculations or estimations have been made regarding how garbage specifically impacts plant and wildlife. One recent study looked at the impact of increasing migration in California after policy changes pushed migrants away from San Diego and into wilderness areas. The *a priori* assumption of this study was that illegal immigrants were causing significant environmental damage in wilderness areas, stating, "Although we cannot definitively prove that immigrants have caused this [environmental] impact, the circumstantial and anecdotal evidence is compelling" (McIntyre and Weeks 2002:393). While the authors acknowledged that Border Patrol also leaves an impact, the main goal of the article was to calculate the environmental impact of each migrant. To do this, they looked at new trail creation, trash, campfires, and wildfires suspected to have been caused by migrants. They found that indeed as more and more migrants traverse wilderness areas, environmental degradation is increasing. In eastern San Diego county, where the study took place, they found that "this damage, manifested in the form of unmanaged footpaths and campsites, illegal campfire, wildfires, and litter, is so widespread that it may be years before the long-term impact is fully understood and even longer before the area recovers" (McIntyre and Weeks 2002:403).

At Cabeza Prieta, there is little evidence of such dramatic foot traffic and migrant discards. Once in a while you might spot a discarded backpack, articles of clothing, or plastic water jug, but these articles are few and far between and are insignificant by comparison to the omnipresent roads and tire tracks. Thus, the perception of migrant garbage degrading the environment at Cabeza Prieta is at times somewhat exaggerated,[3] and a remnant of a standard (and perhaps oversimplified) interpretation of the impact of migration in a context of what Peluso and Wattts (2001:5) term "environmental scarcity." I will explore this phenomenon further in the discussion section of the paper but will first explain the environmental impact of the other culprit in the blame game: United States Border Patrol.

Environmental Impact of Border Patrol

Border Patrol agents have patrolled the American southwestern deserts for more than 100 years. Mounted Watchman began in 1904, many of whom were former Texas Rangers (Annerino 1999). Like Latino migrants, Border Patrol agents also face negative stereotypes; Maril's (2004) ethnography of a Border Patrol unit in Texas offers some humanity to their often difficult plight. Part police officer, part rescuer, agents are often labeled as young, naïve, brutish, and racist (one irony of this is that it is estimated that over 50 percent of Border Patrol agents are of Latino descent) (Pinkerton 2008). The growth of Border Patrol has offered many Latinos and others in the border region a new professional career path, one that offers a decent salary, good benefits, and opportunities for advancement and promotion (see Trac Immigration Reports 2006 for detail on increases in number of Border Patrol agents.). It is not my intent in this article to label individual agents as environmentally-unconcerned, but rather to look at the ways in which the practices and policies of the agency as a whole have caused significant and long-lasting damage to the desert floor.

Contrary to the media portrayal of the environmental havoc wreaked by migrants in the borderlands, during my time on the refuge, it became apparent that many local people and conservation experts attribute much of the environmental damage occurring on the border to Border Patrol.

During fieldwork, I heard the refuge referred to it as the "doormat" of the entrance to the United States. Comments such as the following quotes collected during a town hall meeting in Ajo highlight the tense emotions and vivid perceptions of local residents and illustrate the highly militarized context of the border region:

> "[Border Patrol] are totally trashing the environment."
> "[Hiking on public lands] reminds me of my time in 'Nam."
> Cabeza Prieta is the "battlefield in the war on immigration."

While many local people and refuge visitors I spoke with during fieldwork do see garbage as a problem (and identify garbage as associated with migrants), they also see Border Patrol as a major perpetrator of environmental destruction and as reducing their wilderness experience. An interview respondent summed up the apprehension of many local residents and refuge visitors, explaining "I have some fear that the Border Patrol are out of control, they seem to be unsupervised and act without regard to the rights of citizens." While it is rare to see a migrant or smuggler in the area, images and representatives of the United States military in this region are omnipresent. Gila Bend, the closest town North of Ajo is home to Luke Air Force Base. Pilots conducting their flight training for Iraq and Afghanistan are out on a regular basis. Out on the refuge, military debris is scattered everywhere, including the occasional live ordinance (unexploded weaponry previously used in training exercises). Border Patrol is also visible everywhere, day and night and in the most remote locations. Occasionally, one might see young agents holding large automatic weapons walking alongside the main highway. Border Patrol helicopters fly regularly, and the occasional drone is spotted flying on surveillance. However, even with the multitude of security systems in place, Border Patrol predominately operates out of their vehicles.

The Real ID Act, passed in 2005, gives the Department of Homeland Security (DHS) the authority to waive consideration of environmental impacts of any border-constructions project in the name of national security.[4] Several standard legal requirements have been waived along the border as a result of this legislation, including the National Environmental Policy Act, the Endangered Species Act, the Clean Water Act, the Wilderness Act, the National Historic Preservation Act, the National Wildlife Refuge System Administration Act, the Military Lands Withdrawal Act, and the Sikes Act[5] (Nunez-Neto and Garcia 2007). Included in this legislation is a prohibition against legal challenge; in other words, environmental organizations that have attempted to challenge this act have been denied the opportunity to do so in court. Michael Chertoff, Secretary of Homeland Security from February 2005 until the end of the Bush administration, repeatedly and emphatically adhered to the proclamation that Homeland Security is not bound to follow the rules and regulations of environmental protection in the United States. One refuge employee explained the situation as such: "DHS *is* required to follow refuge rules, however, exigent circumstances exist." Due to the border problems "there have been conflicts. There have been legal conflicts. We have to define exigent."

In terms of an on-the-ground effect on Cabeza Prieta National Wildlife Refuge, these legislative developments have affected environmental degradation in two primary ways. The first is that the construction of the vehicle barrier fence on the south side of the refuge was not monitored or bound to follow typical environmental protection and mitigation procedures. SunCorp, the contractor building the vehicle barrier along the border, was under no supervision or subject to any environmental regulation. Large earthmoving machines have dotted the southern boundary of the refuge for years.

The second effect on Cabeza Prieta is that Border Patrol Agents are not required to follow standard federal policies or procedures to minimize their environmental impact on the refuge. In short, Homeland Security trumps environmental protection in every scenario, in every location, and at all times. Chertof's stance on avoiding environmental regulations was upheld by the Supreme Court in 2008. Without the ability to challenge the Real ID Act in court, it is unlikely that this situation will change anytime in the near future.

New Border Patrol agents working within the boundaries of Cabeza Prieta (and adjacent Organ Pipe National Monument) receive minimal training on how to reduce their environmental impact. I was able to attend a Border Patrol environmental "training" seminar in winter 2009. The training consisted of two short videos. The first video, *Protecting Our Environment,* was produced by the Tucson Sector Public Information Office for Border Patrol and discussed the issue of endangered species in the area, and how, because of the cryptogamic soil, tire tracks in the desert can last 10 to 50 years. After I watched it, the agent showing the video revealed that they actually don't show this video to new agents any more. They feel it is more important to get new agents out into the field immediately and don't have time for electives like environmental training (presumably, then, he was showing it to me because of my stated interest in environmental issues). The second video, *Patrolling in a Desert Ecosystem,* primarily focused on teaching agents how to appease environmental organizations that often are unhappy with the presence of Border Patrol in fragile areas. To paraphrase, the voice-over explained "there are groups out there that want Border Patrol not to work in delicate areas. So, do your job, but manage your impacts, or there will be resistance from some of these groups."

During this training, I heard what would become a standard Border Patrol mantra when confronted by adversaries. They justify their lack of environmental training on the fact that their agency is currently experiencing "growing pains." In other words, because the agency is growing so fast, they are unable to train their new agents on environmental issues. There is some truth to this—since September 11, 2001, Border Patrol has grown significantly, though this growth has been steady for the past few decades, and more rapid growth has been occurring since the 1990s. But it's also true that as long as there is no institutional mandate that environmental protection becomes a priority, there will likely not be much in the way of training for individual agents.

Illegal Roads

The increase in the number of Border Patrol agents logically leads to an increase in the overall environmental footprint of activities. According to one conservation expert, "The vast majority of new roads from 2002 onward are from Border Patrol." But according to a Border Patrol public relations officer I interviewed, "Roads are usually made by illegal aliens." This demonstrates stark differences in the agency-level perceptions of environmental degradation, yet it is difficult to empirically demonstrate since statistics on the number of on-foot border-crossers compared to vehicle border-crossers are not kept.

While finger-pointing may be the norm in the local debate on environmental degradation, no data exist to identify the primary causes of new road creation in and around the refuge. Thus, while it became glaringly evident early on in my fieldwork that the use of administrative roads and off-road driving was having a significant impact on the environmental health of Cabeza Prieta, it was not clear who was most responsible for the illegal roads. In the cat-and-mouse game of apprehension on the border, were the smugglers creating new roads with Border Patrol in pursuit, or was Border Patrol driving off-road simply because they could, because they were not bound to follow environmental regulations?

The absence of data on user-groups and their environmental impact in Cabeza Prieta is a major gap in information and is not benign. For Border Patrol, the public perception that "illegals" cause the greatest environmental harm diminishes the amount of public pressure on them to reduce the environmental impacts of their work. According to a Border Patrol spokesperson, "The best way Border Patrol can help protect the environment is by stopping [migrants/smugglers] at the border." This is a bold statement and suggests that the damage caused by increased vehicular monitoring of the border is a worthy cause, regardless of the impact of their work. Without careful monitoring of Border Patrol activities (a sanctioned activity), monitoring the effects of unsanctioned activity like migration and smuggling cannot be quantified. To address this lack of data, and utilizing publicly available aerial photography and geographic information system software, I examined the issue of road creation and vehicular damage across the refuge, and I present a very brief summary of the results here.

A vast system of administrative trails and dirt roads stretches across the refuge. Many of these roads existed prior to the government's designation of a wildlife refuge or wilderness. Some roads are historic (such as the infamous *Camino del Diablo* [Highway of the Devil]), and others are more recent creations, built to aid in the distribution of supplementary water to Bighorn Sheep and other wildlife. No one knows for certain how often these pre-existing roads are traveled, or if they tend to be utilized more often by smugglers or Border Patrol agents or off-road recreational users. What is certain is that the environmental damage caused from vehicular traffic is significant and long-lasting.

The overall growth in roads in the refuge is more than just an aesthetic nuisance. Vehicular traffic bothers wildlife and destroys flora. Growth in new roads creates an ever-escalating problem; as Border Patrol agents drive off road in pursuit of border crossers, they create new pathways that can be used by smugglers, creating an exponentially increasing network of vehicle routes. Further, as more and more traffic drives on the refuge, the width of individual roads continues to increase. Finally, since the roads run along valley floors, when there is rainfall, the run-off is trapped in the deep rivets. As a result, on rare occasion when there is a significant rain event, illegal roads reduce the area of the green band of vegetation available for wildlife. This directly affects endangered species like the Sonoran Pronghorn.

To examine the issue of illegal road creation, I analyzed publicly available United States Geological Society Digital Quarter-Quad (USGS DOQQ) imagery for the year 2005. The images were processed and analyzed through the use of the geographic information system ArcGIS 9.3. The project should be seen as a preliminary endeavor in an effort to quantitatively assess road length increases in the refuge. In my analysis, I focused on a small portion in the southeast corner of the refuge, an area known for vehicular traffic and directly adjacent to the border. I created a polygon along the southern border of the refuge extending approximately five miles north. I reviewed 35 quarter quads, drew in visible roads and trails, and then calculated the total length of visible roads. Within my study area, there were 31.6 miles of official, documented roads, but at least 95.7 miles of roads visible in the images. Thus, there were at least 64 miles of illegal or undocumented roads in this corner of the refuge (and note that I erred on the side of caution in my analysis, only drawing in roads that were clearly visible).

The results of my remote-sensing study combined with interview and observational data indicate that roads are indeed increasing within the refuge. This is especially true when looking at horizontal expansion of individual roads. As the roads are used more and more and the sand becomes more difficult to navigate, people tend to drive on the outside edges of the roads, and over time this creates wider roads. In addition to suggesting an increase in roads, the remote-sensing project provides insight into just how obscure the problem of road increases are—my study was the first to actually document the extent of the illegal roads (though a more extensive study was completed in 2010 by refuge staff, described below). While I am not able to ascertain which group is the primary perpetrator of the illegal roads utilizing remote data, interview-gathered anecdotal data suggests that Border Patrol is certainly the most visible group creating new roadways.

In the time since my own research was conducted, new data was released related to the impact of roads. According to one refuge manager at Cabeza Prieta, "We have mapped over 3,100 miles of vehicle trails in this area. This is approximately 70 times the number of trails that were present at the time of wilderness designation in 1990" (United States Fish and Wildlife Service 2010:10). Unfortunately, the extent of the impact of illegal road growth remains uncertain. According to the same refuge manager, "We conservatively estimate there are at least three times as many frequently used roads present on the refuge than at the time of wilderness designation; this does not include the lesser used trails that wind through virtually every valley within the refuge" (United States Fish and Wildlife Service 2010:4).

Even with this new data, documentation on the extent of environmental degradation is sorely lacking. One important next step for both land managers and external researchers will be to gather more data on the extent of road damage through the use of remote sensing and GIS, as well as subsequent ground-truthing. Further, it will be important to develop techniques to identify the vehicles, through the use of technology such as remote motion-sensor cameras.

The Blame Game: Shifting Perceptions

In this final section of the paper, I present an analysis of the varied and conflicting perceptions of the causes of environmental damage on the refuge. I argue that the divergent ways in which people assign responsibility for environmental damage are embedded in broader social and political systems. The analysis of environmental degradation at Cabeza Prieta leads to an important question: why has the focus on trash received more attention than the issue of illegal roads? I argue that the perception of the Mexican degradation of the border is connected to deeply held stereotypes and beliefs about "dirty Mexicans." As other observers of the border have noted, the "notion of the 'dirty Mexican' runs deep in United States culture" (Sundberg 2008:877). Mainstream America now tends to view Latinos via "potentially threatening images" (Santa Ana 2002:8). The image of the dirty Mexican is emblematic of a broader shift in American society towards an increasingly negative perception of Mexican immigrants. As Santa Ana (2002:8) argues, examining Latino metaphors, such as the "Mexican as invader" or "Mexican as degrader" offers "a window on the ways that Americans frame their domestic worldview and on their underlying political and social values." The following section offers insight into how this threatening image has developed in the environmental realm.

In the popular media, the environmental focus on the border tends to highlight the impact that "illegal aliens" have on the landscape. In both local and national media, "trash becomes the focus of the articles, rather than border and immigration policies, and undocumented immigrants are framed as those who 'trash'America" (Sundberg 2008:874). Dramatic stories, filled with images of discarded water bottles, clothing and other personal belongings left behind at unsanitary camping spots, tell and re-tell a common story—the Mexicans are invading our pristine wilderness areas and damaging nature. With upswings in drug and human smuggling, it has been asserted that "illegal immigrants crossing from Mexico have created a vast system of 'major highways' across wilderness" (Kenworthy 2006:). But the "trash-talk" discourse present at the border has as much to do with issues of social identity as with environmental degradation (Sundberg 2008). As Leo R. Chavez (2001) has pointed out, images of Mexican immigration into the United States tend to be alarmist in nature, an observation that is supported by my own analysis of the perceptions of environmental degradation on the United States-Mexico border.

What has resulted is an environmental discourse along the United States-Mexico border that links environmental degradation in the region as threatening and dangerous to the American people and the American nation (Hill 2006; Sunderberg and Kaserman 2007). This is caused in part simply by the sheer number of migrants, which invokes neo-Malthusian fears of over population (Urban 2008). But the metaphor of the dirty Mexican contributes to a fear-based environment. Other metaphors employed to dehumanize Mexican-origin migrants range from Mexicans as an invasive species and immigrants as floods or deluges. (For a fascinating analysis of Mexican metaphors see Santa Ana 2002.)

Sundberg (2008) argues that emphasis on Mexican-origin trash at the border is a part of a greater process of the "geographies of exclusion." In other words, the trash is out of place because it is Mexican. The issue is not so much that there is trash, but rather where that trash comes from. As previous anthropological studies have shown, what constitutes pollution is subjective (Douglas 1966).

Sundberg's geographies of exclusion can be contrasted with Peluso's (1996) "ethics of access." Peluso (1996:515) explains, "The ethic of access is driven by more than economics or subsistence rights and serves social, political, and ritual purposes as well, representing kinship, power relations, and ritual harmony. At Cabeza Prieta, the politics of access are linked to issues of who belongs in a wilderness area and who doesn't belong. For example, while the two primary groups accused of degrading the refuge are migrants/smugglers and Border Patrol, there are actually a number of different stakeholders present at any given moment in the conservation site, including:

- migrants/smugglers
- Border Patrol and related Homeland Security activities
- staff and volunteers of Cabeza Prieta
- recreational visitors
- United States Air Force
- Arizona game and fish

Each of the above groups is allowed varying degrees of entrance to the refuge and permitted varying levels of activity. Entrance and type of allowable activity is primarily monitored by the issuance of permits; any documented, lawful visitor to the refuge must obtain a permit to gain permission to enter the area. This creates a process whereby some user groups are permitted (both legally and socially) in the refuge, and some groups are not permitted within the refuge. When we look at the bigger picture of access, a more broadly defined group of stakeholders, or "user-groups," emerges.[6] Each user group has some sort of environmental impact on the refuge. For example, recreational visitors travel in four-wheel drive vehicles, and there is a large amount of military debris scattered across the refuge from military training exercises. Even conservation activities led by refuge staff have an environmental footprint—for example, flying and trucking in supplementary water for wildlife. However, the focus on environmental degradation tends to become blurred by the social, political, and economic tensions inherent in the immigration debate, and the blame game narrows to simply Border Patrol versus Mexican migrant, even though each of the above listed groups has some environmental impact.

Drawing upon the concept of an ethics of access at Cabeza Prieta is one lens through which to understand how other political, social, and ethnic structures are always present in environmental preservation projects, especially in a border conservation context. Who has the right to be present inside a conservation site? Nature preservation has always been in a process of denning who belongs in a protected area (e.g., Cronon 1996; Nash 1973). Further, issues of race and ethnicity are always present in conflicts over access, even if not overtly. To expand on this idea, I argue that issues of access and allocation present in struggles over border protected areas are inherently tied to issues of national security and national belonging. Stoller (2008:193) argues, "Imperial formations are denned by racialized relations of allocations and appropriations." In the United States-Mexico border context, then, it becomes impossible to separate issues of race (and class), national security, and environmental protection, and through a perpetuation of the belief that the refuge is under threat of a Mexican invasion, Border Patrol's presence in the region tends to be less critiqued or questioned. In a sense, their legitimacy is reinforced (and environmental footprint justified as a direct result of the immigrant threat).

Nevins (2002) argues that the state actually helped to create the concept of the "illegal" Mexican through the construction of fences and walls and the buildup of Border Patrol; in other words, the state helped to create the very problem (i.e., illegal immigration) that it is now fighting against. Nevins

(2002:13) explains, "In one sense, [immigration policy], by trying to limit unauthorized entries in the United States, is an effort to eliminate boundary-related illegality. Yet, at the same time, the operation helps to construct and perpetuate illegality." Along the same vein, the state helped to create the problem of environmental degradation along the border (through shifts in immigration policy), and now even greater resources must be leveraged to protect these fragile areas.

Ultimately, the blame game creates a type of unintended feedback mechanism that is often present in border policies and contributes to a persistent escalation in damage and tension. The feedback and escalation process is not unique to environmentally-centered issues. Escalation is fueled by the "primacy of image and symbols for state actors engaged in border management" (Andreas 2009:9). In other words, images of trash and perceptions of dirty Mexicans propel border policies in specific and predictable directions. The emphasis on trash and the narrowing of the blame game is indicative of a shift towards a stronger linking of environmental issues and security concerns in the border region, a process termed "ecological security" (Homer-Dixon 1999) or "green security" (Peluso and Watts 2001).

An over-focus on trash as the main type of environmental degradation can also be understood as a part of the process of what has been termed the "greening of hate" (Hartmann 2004). The "Latino Threat Narrative" (Chavez 2008) in this case, has specific environmental undertones. According to Chavez (2008:3), the Latino Threat Narrative "is part of a grand tradition of alarmist discourse about immigrants and their perceived negative impacts on society." It is easy to see how this alarmist discourse can manifest itself within the context of environmental concerns.

Conclusion

In this paper, I have focused on the perceived negative environmental impacts of undocumented immigration. I find that, ultimately, in the context of the United States-Mexico border region, environmental concerns cannot be separated from social constructs such as race and ethnicity. In my initial research, I found that environmental degradation on the border tends to be divided into two categories: the illegal migrant/ smuggler desecration of the land or the Border Patrol driven degradation. In actuality, the causes of environmental harm are not so linear, and multiple actors contribute to degradation in a number of different ways. While the causes of degradation are varied, I have found vehicular traffic to be the most prevalent and damaging activity.

Ultimately, the conflicting perceptions of the causes of environmental damage are grounded in subtle but powerful social beliefs about who has the right to be physically present in the refuge. Perceptions of environmental degradation are influenced by the broader social, racial, and political tensions present on the United States-Mexico border. An illumination of some of the cultural and political aspects of nature conservation can help "call attention to the broader political stakes of environmental politics" (Kosek 2006:21). In other words, studies of environmental issues, especially in contentious places like the United States-Mexico border can inform us about broader societal tensions and provide a unique lens through which to analyze issues of race and ethnicity in America A better understanding of how such environmental perceptions are formulated can improve information channels for policy decision making, both in terms of environmental policy and immigration policy.

Notes

1. Wackenhut is the name of the company hired to transport apprehended individuals back to the border.
2. Wildlife Refuges are administered through the United States Fish and Wildlife, under the United States Department of Interior.
3. Note that impacts are in a constant state of flux and change, as popular crossing routes shift in response to economic factors (like the recession) and to border patrol clampdowns. Other protected areas along the border that experience more migrant and smuggler traffic have seen more of an impact from garbage. Buenos Aires National Wildlife refuge has received some media coverage of the problem (e.g., Lacey 2010).
4. "The primary intention of this law was to create a nation-wide standard for obtaining driver's licensees (i.e., state issued identification). The environmental release was attached as a rider.
5. The Sikes Act requires the Department of Defense to initiate programs of nature conservation on military lands.
6. "I prefer to employ the term "user-group', rather than stakeholder because, in the context of Cabeza Prieta, so many of the interested groups are transitory and in motion and not interested in any long-term investment or resource extraction–they simply want to "use" the land.

References

Andreas, Peter 2009 Border Games: Policing the Untied States-Mexico Divide. 2nd ed. Ithaca, N.Y.: Cornell University Press.

Annerino. John 1999 Dead in Their Tracks. New York: Four Walls Eight Windows.

Associated Press 2000 "Chertoff defends southwest border fence." In USA Today, Posted 10/1 /2007. http://www.usatoday.com/news/washington/2007-10-01-chertoff-fence_N.htm (December 7, 2011).

Brechin, Steven R. 2003 Contested Nature: Promoting International Biodiversity With Social Justice in the 21st Century. Albany: State University of New York Press.

Chavez, Leo 2001 Covering Immigration: Popular Images and the Politics of the Nation. Berkeley: University of California Press. 2008 The Latino Threat: Constructing Immigrants, Citizens, and the Nation. Stanford, Calif.: Stanford University Press.

Clarke, Chris 2008 The Battered Border: Immigration Policy Sacrifices Arizona's Wildernesss. URL:<http://www.earthisland.org/journal/index.php/elj/article/the_battered_border> (October 20, 2010).

Cornelius, Wayne 2001 Death at the Border: Efficacy and Unintended Consequences of United States Immigration Control Policy. Population and Development Review 27(4):661–685.

Cronon, William 1996 Uncommon Ground: Rethinking the Human Place in Nature. New York: W.W. Norton and Company.

Donnan, Hastings, and Thomas Wilson 2010 Borderlands: Ethnographic Approaches to Security, Power, and Identity. Lanham, Md.: University Press of American, Inc.

Douglas, Mary 1966 Purity and Danger: An Analysis of the Concepts of Pollution and Taboo. London, United Kingdom: Routledge and Regan Paul.

Ferguson, Kathryn, Norma Price, and Ted Parks 2010 Crossing With the Virgin: Stories From The Migrant Trail. Tucson: University of Arizona Press.

Fox News 2008 Groups Struggle to Clean Up Mess Illegal Immigrants Leave Behind. URL:<http.//www.foxnews.com/ story/0,2933,354398,00.html> (December 7, 2011).

Gauthier, Melissa 2010 Researching the Border's Economic Underworld: The "Fayuca hormiga" in the United States-Mexico Borderlands. *In* Borderlands: Ethnographic Approaches to Security, Power, and Identity. Hastings Donnan and Thomas M. Wilson, eds. Pp. 53–72. Lanham, Md.: University Press of America, Inc.

Hartmann. Betsy 2004 Conserving Racism: The Greening of Hate at Home and Abroad. URL:<http://popdev.harnpshire.edu/sites/popdevfiles/ uploads/dt/DifferenTakes_27.pdt> (December 7 2011).

Hill, Sarah 2006 Purity and Danger on the United States-Mexico Border 1991–1994. South Atlantic Quarterly 105(4):777–800.

Homer-Dixon, Thomas F. 1999 Environment, Scarcity, and Violence. Princeton, N.J.: Princeton University Press.

Kenworthy, Tom 2006 New Outlaws Plague Arizona's Desert Refuges. USA Today (McLean, Va.). August 22:2. URL>:http://www.usatoday.com/ news/nation/2006-08-22-parks-borders x.htm. (December 7 2011).

Kosek, Jake 2006 Understories: The Political Life of Forests in Northern New Mexico. Durham, N.C.: Duke University Press.

Kraly, Ellen Percy 1995 United States Immigration and the Environment: Scientific Research and Analytic Issues. Department of Geography, Colgate University: United States Commission on Immigration Reform.

Lacey, Marc 2010 Water Drops for Migrants: Kindness, or Offense? URL:<http://www.nytimes.com/2010/09/27/us/27water.html?_ r=3&ref=global-home> (December 7 2011)).

Maril, Robert Lee 2004 Patrolling Chaos: The United States Border Patrol in Deep South Texas. Lubbock: Texas Tech University Press.

Martinez, Oscar J. 2006 Troublesome Border. Tucson: University of Arizona Press.

McCombs, Brady 2009 Death Count Rises With Border Restrictions Officials: Crossers Now Trek Farther to Dodge Security. Arizona Daily Star. www.AZstarnet.com. , May 17th, 2009. URL>:http://azstamet.com/news/local/border/article_5eafc47d-9751-54db-80f6-b76fdl95df42.html .(December 7 2011)

Mcintyre, David, and John Weeks 2002 Environmental Impacts of Illegal Immigration on the Cleveland National Forest in California. The Professional Geographer 54(3):392–405.

Nash, Roderick 1973 Wilderness and the American Mind. New Haven, Conn.: Yale University Press.

Nealson, Christina 2009 Garbage Grows Well on the Border. High Country News (Paonia, Colo.), April 30 2009. URL> http://www.hcn.org/wotr/garbage-grows-well-on-the-border>. (December 7 2011).

Nevins, Joseph 2002 Operation Gatekeeper: The Rise of the "Illegal Alien" and the Making of the United States-Mexico Boundary. New York: Routledge.

Nuñez-Neto, Blas, and Michael John Garcia 2007 Border Security: Barriers Along the United States International Border. Washington, D.C.: Congressional Research Service.

Oglethorpe, Judy, Jenny Ericson, Richard E. Bilsborrow, and Janet Edmond 2007 People on the Move: Reducing the Impact of Human Migration on Biodiversity. Washington, D.C.: World Wildlife Fund and Conservation International Foundation.

Peluso, Nancy 1996 Fruit Trees and Family Trees in an Anthropogenic Forest: Ethics of Access, Property Zones, and Environmental Change in Indonesia. Comparative Studies in Society and History 38(3):510–548.

Peluso, Nancy, and Michael Watts 2001 Violent Environments. Ithaca, N.Y.: Cornell University Press

Pinkerton, James 2008 Hispanics Bolster Border Patrol They Now Hold 52 Percent of Agents' Jobs. URL:<http://www.chron.com/disp/story.mpl/front/6184818.html> (May 15 2011)).

Pirages, Dennis, and Theresa Manley DeGeest 2004 Ecological Security: An Evolutionary Perspective on Globalization. Lanham, Md.: Rowman and Littlefield

Public Employees for Environmental Responsibility 2008 America's Ten Most Imperiled Wildlife Refuges. URL:<http://www.peer.org/news/news_id.php?row_id=1052> (April 28, 2011).

Pulido, Laura 1996 Environmentalism and Economic Justice: Two Chicano Struggles in the Southwest. Tucson: University of Arizona Press.

Regan, Margaret 2010 The Death of Josseline. Boston, Mass.: Beacon Press.

Santa Ana, Otto 2002 Brown Tide Rising: Metaphors of Latinos in Contemporary American Public Discourse. Austin: University of Texas Press.

Slown, John 2003 Taking Refuge. URL:<http://www.fws.gov/southwest/refuges/plan/refuges.pdf> (April 28, 2011).

Stoller, Ann Laura 2008 Imperial Debris: Reflections on Ruins and Ruination. Cultural Anthropology 23(2):191–219.

Sundberg, Juanita 2008 "Trash-Talk" and the Production of Quotidian Geopolitical Boundaries in the United States-Mexico Borderlands. Social and Cultural Geography 9(8):871–890.

Sundberg, Juanita, and Bonnie Kaserman 2007 Cactus Carvings and Desert Defecations: Embodying Representations of Border Crossings in Protected Areas on the Mexico-United States Border. Environment and Planning D: Society and Space 25(4):727–744.

Trac Immigration Reports 2006 Border Patrol Expands But Growth Rate After 9/11 Much Less Than Before: Division Between North/South Border Little Changed. URL:<http://trac.syr.edu/immigration/reports/143> (May 15 2011)).

United States Fish and Wildlife Service 2010 Compatibility Determination Construction of a Communication Tower Cabeza Prieta National Wildlife Refuge. URL:<http:// www.fws.gov/southwest/refuges/arizona/cabeza/docs/CPNWR.FinalCD.SBInetTowerl89.pdf> (April 28, 2011).

United States Geological Society (USGS) 2007 Map of Protected Areas on the United States-Mexico Border. URL:<http://borderhealth.cr.usgs.gov/rnaps/fed_landsmap.pdf> (October 29, 2009).

United States Government Accountability Office (GAO) 2006 Illegal Immigration: Border-Crossing Deaths Have Doubled Since 1995: Border Patrol's Efforts to Prevent Deaths Have Not Been Fully Evaluated. Washington, D.C.: United States Government Accountability Office.

Urban, Jessica LeAnn 2008 Nation, Immigration, and Environmental Security. New York: Palgrave MacMillan.

Wilderness Society 2008 Cabeza Prieta National Wildlife Refuge: A Landscape Under Assault. Washington, D.C.: Wilderness Society.

Section 3

RESPONSE

Chapter 9

Addressing State Failure

By Stephen D. Krasner and Carlos Pascual

The Danger of Failed States

In today's increasingly interconnected world, weak and failed states pose an acute risk to U.S. and global security. Indeed, they present one of the most important foreign policy challenges of the contemporary era. States are most vulnerable to collapse in the time immediately before, during, and after conflict. When chaos prevails, terrorism, narcotics trade, weapons proliferation, and other forms of organized crime can flourish. Left in dire straits, subject to depredation, and denied access to basic services, people become susceptible to the exhortations of demagogues and hatemongers. It was in such circumstances that in 2001 one of the poorest countries in the world, Afghanistan, became the base for the deadliest attack ever on the U.S. homeland, graphically and tragically illustrating that the problems of other countries often do not affect them alone.

The international community is not, however, adequately organized to deal with governance failures. The United States and the rest of the world need to develop the tools to both prevent conflict and manage its aftermath when it does occur. Such efforts will entail not just peacekeeping measures, but also influencing the choices that troubled countries make about their economies, their political systems, the rule of law, and their internal security. Weak countries are unable to take advantage of the global economy not just because of a lack of resources, but also because they lack strong, capable institutions. To promote sustainable peace, Washington and its partners must thus commit to making long-term investments of money, energy, and expertise.

The United States is moving in the right direction. Following a decision of the National Security Council in the spring of 2004, the Bush administration created a new office within the State Department: the Office of the Coordinator for Reconstruction and Stabilization. S/CRS will help lead and coordinate joint operations across agencies to respond effectively to evolving crises around the world, in concert with the international community. The White House has requested $124.1 million from Congress to finance the first phase of the new office and the programs it will support. The price for building a rapid-response capability is small. It is miniscule compared to the cost of ignoring the threats posed by failed states.

Conflict prevention must become a routine element of policymaking. Leaders in Congress, the administration, and the nongovernmental community must continue to devote their energies to stabilizing the vulnerable regions of the world. The U.S. government must be able to anticipate potential problems quickly and effectively so that they can be managed before they develop into full conflagrations. There is always the risk that prevention in any given situation may fail, and that must be accepted—both by senior policymakers and by the entire government. Crises will inevitably occur, but if they are the United States' only impetus for response, there will be less chance of success.

Learning and Acting

The many U.S. agencies and departments devoted to peacekeeping and development have learned important lessons about effective conflict management—in both prevention and reconstruction. It has become a truism that preventing armed conflict is preferable to resolving differences through force: prevention saves both lives and money. Yet the international community often does not fully appreciate the complexity and difficulty of preventing belligerent parties or criminal militias from going to war. Countries involved in simmering conflicts almost universally lack the requirements for successful stabilization: an effective indigenous leadership that puts national interests over its own; state institutions that are capable, legitimate, and credible; and a citizenry that actively participates in its own governance. Moreover, the traditional tools of diplomacy—démarches, treaties, dialogue—often have limited sway over actors unconcerned about their international image or legitimacy.

Further complicating matters, modern conflicts are far more likely to be internal, civil matters than to be clashes between opposing countries. "State death" as a result of external invasion, common before World War II, has almost disappeared since 1945. The lack of good governance in weak states means they often do not have the ability to deal with disaffected or criminal groups within their own borders. Recent scholarship suggests that civil strife is no more likely in ethnically or religiously divided countries than it is in homogeneous ones. Internal discord is more likely to arise in countries suffering from poverty, a highly unequal income distribution, recent decolonization, weak institutions, ineffective police and counterinsurgency forces, and difficult terrain—conditions that allow small guerrilla bands to thrive. Valuable raw materials, such as diamonds or oil, also tend to spark conflict among competitors who want to seize control of the wealth. Warring groups generally have easy access to weapons and may even control territory, giving them a base for launching attacks on the state, its citizens, or its neighbors. Other nonstate actors, including transnational terrorist organizations, can also take root in such environments, posing a threat to global security.

These elements of state weakness constitute structural threats akin to dead leaves that accumulate in a forest. No one knows what spark will ignite them, or when. Over the long run, the only real way to create lasting peace is to promote better governance. The United States and the international community have increasingly made governance a focal point of development. The U.S. Agency for International Development (USAID) now spends about $1 billion a year on governance assistance. In 2002, in Monterrey, Mexico, international leaders agreed that good governance in developing countries should be rewarded with more resources. The Bush administration's Millennium Challenge Account will provide higher levels of aid to countries that have demonstrated political and economic reform. And the World Bank has increasingly introduced governance criteria into its lending decisions. But governance assistance designed to accomplish long-term change may not have much impact on countries where civil strife is imminent. Untargeted resources channeled into tense environments could even wind up making the conflict worse.

Anticipating, averting, and responding to conflict require a greater, more comprehensive level of planning and organization—which is precisely the mission of S/CRS. The coordinator reports directly to the secretary of state and has a staff drawn from functional bureaus in the State Department, as well as from USAID, the CIA, the military's Joint Staff, the Office of the Secretary of Defense, and others. Within the U.S. government, S/CRS is building civilian capacity to plan and coordinate stabilization and reconstruction efforts. S/CRS and its interagency partners also draw on expertise from nongovernmental organizations, think tanks, private firms, and universities. The U.S. government will use these resources to encourage and coordinate activities with other governments and international organizations. To focus attention on conflict prevention, S/CRS is also partnering with other offices, including those of the director of policy planning, the assistant secretary for intelligence and research, and the assistant secretaries of state in the regional and functional bureaus. The undersecretary for political affairs has strongly supported this engagement.

The United States cannot address every potential conflict, but it can determine which ones present the worst threat and which can be best helped by international attention. Every six months, the National Intelligence Council currently produces a list, compiled from classified and unclassified sources, of the countries at the greatest risk of instability. The list allows agencies across the U.S. government to assess state fragility, chart any changes in status, and develop strategies in case conflicts emerge. Regional bureaus use this crucial information to decide where the United States should focus its attention, personnel, and resources. S/CRS is working to incorporate the expertise of other government agencies and, increasingly, international counterparts into these assessments, to make them more comprehensive and reliable.

Each conflict is unique, and the U.S. response must be tailored to specific situations, but better understanding the broader factors that influence conflicts will help the government recognize warning signs and craft appropriate strategies. To this end, S/CRS is sponsoring expert roundtables on specific countries at risk of instability to generate new policy ideas on prevention. The office is also helping civilian organizations incorporate proven military analytic tools for conflict management and prevention:

common planning templates, simulated exercises to test new ideas, and checklists of actions that help transform postwar environments by addressing the factors that drive conflict.

The better the U.S. government operates internally, the more effective it will be in forging partnerships with other important international and nongovernmental actors. As the United States develops a single approach to conflict management, it can improve its coordination with organizations such as the UN, the European Union, and NATO, as well as with nongovernmental organizations and key bilateral partners. The United Kingdom, Canada, Germany, and France are all considering or making their own structural changes to anticipate and manage foreign conflicts. UN reforms on "peace building" provide an opportunity to develop a multilateral context within which these nations can cooperate. There is an unprecedented acknowledgment throughout the world of the need to work together to prevent or manage state failure—and that not doing so would threaten international security.

Understanding Transformation

Even with investments in prevention, violence will still erupt and demand attention. To manage postconflict engagements effectively, the international community must understand the nature of the changes that need to occur to transform the affected states into ones in which further conflict will be unlikely. The international community must also commit the necessary resources. Although real life is more complex than any model of postconflict transition, dividing the transition into four phases can help guide future responses. These phases are not necessarily sequential—some may proceed concurrently, and they may not progress consistently.

The first phase is stabilization—the stage that generally garners the most international attention. Stabilization requires taking immediate action: enforcing order, feeding people, restarting basic services, initiating a political transition process, generating local employment, and reintegrating returning refugees and internally displaced persons. The international community often plays a dominant role in this phase, directly delivering services rather than building local capabilities, in order to avert chaos. Even if necessary in the short run, such efforts cannot be sustained indefinitely. Hence, outsiders must at some point stop "doing" and begin "enabling" local involvement and ownership. The faster the international community responds initially, the easier it will be to make sure this transition takes place. Stabilization must lead to conditions that help economic, political, and social development—perhaps most crucially by engendering local leadership and providing incentives and means for local parties to take action. This first period is critical to a country's future success.

Second, the conflict's root causes must be addressed. Such factors can include corruption, collapsed economic systems, political exclusion, or the private exploitation of public resources. If left unchecked, these issues will continue to stymie progress and economic growth in a postconflict setting, particularly in former authoritarian states. Yet the very act of unraveling the evils of the past can be destabilizing. Shutting down bankrupt state enterprises generates unemployment. Opening up the political system challenges vested interests and political elites. Holding former oppressors accountable at the national level can strain barely functioning institutions. The international community must thus make sure that there is political consensus for reform, and that social safety nets, particularly for the unemployed and pensioners, are developed to prevent tensions from undermining the prospects for success.

Next is the creation of the laws and institutions of a market democracy—or fostering the "supply side" of governance. Many of the features of good governance that advanced states take for granted must be built anew or reinvented in countries that have just emerged from war. These elements include markets, tax systems, banks, and regulatory policies to make an economy function; constitutions, political structures, parties, and electoral processes to underpin participatory politics; and laws, courts, police, and penal systems to create the rule of law. Out of all four phases, this one may be the hardest since it entails defining a new polity. In many cases, the shift is as extreme as replacing authoritarianism with openness, freedom, and competition. Such changes are radical, affecting every aspect of life. They must be woven into the fabric of society. As Secretary of State Condoleezza Rice put it in a speech earlier this year, democracy is a process, not an event.

Finally, the "demand side" of politics, essential for accountability, must be established. Only if there is effective demand from the governed can democratic institutions be sustained. To generate this demand, civil society must evolve—communities need to develop as constituencies that call for political attention for their needs. Rebuilding social cohesion and public confidence in the government will, of course, take time: countries coming out of conflict are often deeply divided. Independent media are also crucial to this process—to help educate the population, reflect public opinion, and keep government officials honest. The free flow of information is essential to a flourishing democracy.

Outsiders can assist, advise, and provide incentives for locals. But ultimately, those within a society must define their own future. Even then, the path will not be easy. The example of the Soviet Union is instructive. Its collapse was generally peaceful; it boasted a working infrastructure, significant resources, and high levels of education—yet 14 years later, the future of decent governance and democracy is still in doubt in many of its former constituent states. The challenge will be much greater in countries wracked by poverty, illness, and illiteracy in the aftermath of war.

If the United States and the rest of the international community do not understand and plan for long-term transitions in advance, the chances of success diminish. Elections are generally not an endpoint, but only one of many necessary steps to build local legitimacy. International actors must be able to shift out of crisis-response mode to supply normal support mechanisms without losing attention or commitment. There will inevitably be international disillusionment with slow progress as countries take control of their own reconstruction and struggle to use the aid they are given. At this point, leadership from the White House and Congress will be key. Decision-makers must maintain a long-term perspective and remember that, although the complexity of transition is frustrating, cutting off funding only increases the likelihood that a country may lapse into chaos.

Innovation

If prevention fails, the fundamental challenge of S/CRS is to make sure that the United States is ready and able to manage all four postconflict stages. No single government office can take on this sizable task alone. But the nation now has a focal point to lead and coordinate rebuilding efforts with other U.S. agencies and with the rest of the world. The Bush administration is using S/CRS to develop a new set of tools for conflict response. These tools fall into six basic categories.

First, the coordinator's office is developing both the framework and the capability to plan for stabilization and reconstruction. The framework will ensure that the government's postconflict goals are realistic and linked to the resources needed to achieve them. If the goals and resources do not match, senior policymakers will have to make tough choices about committing additional resources or revising expectations. Setting out at the beginning a catalog of essential tasks will force planners to address both immediate needs and the eventual transition from international to indigenous leadership and sustainable development. In short, the framework will require the United States to develop a long-term perspective from the outset.

Second, S/CRS is making sure that the government is ready to move rapidly to help countries in the aftermath of conflicts. The United States must be better able to provide central leadership and management quickly. It must be ready to use diplomacy to diffuse local tensions, even as it deploys civilians and the military to establish order and initiate, manage, and oversee assistance and rebuilding. To create this capability, the coordinator will need a staff of 80 people and an active response corps of 100 diplomats who would spearhead diplomacy or augment local U.S. embassies in an emergency. Advance civilian teams of diplomats and specialists from other agencies would deploy to the field to foster reconciliation, address public safety, coordinate with international groups, and help broker peace agreements and arrangements for political transitions. In addition to these activities, the teams would constantly take stock of on-the-ground realities and use this information for long-term planning.

To support the larger and longer-term program requirements, the coordinator's office is assessing and filling gaps across government agencies in contracts and more informal arrangements with organizations that specialize in various aspects of stabilization and reconstruction: mobilizing international civilian police, training indigenous police, developing systems of justice, providing fiscal and monetary advice, stimulating the private sector, and supporting civil society. S/CRS is also assessing the feasibility of a civilian reserve corps that could tap individuals with key skills. The goal is to organize all of these resources so that they can mobilize quickly and efficiently after a conflict to fill all the needed functions and skills.

Applying these resources will, of course, take funding. The third tool S/CRS requires is a $100 million conflict-response fund and more flexible spending authority. The Bush administration has proposed that Congress create such a fund, and Senators Richard Lugar (R-Ind.) and Joseph Biden (D-Del.) have championed legislation to this effect. An available supply of money would allow the government to respond faster, creating a greater chance to generate positive momentum, reconcile warring parties, and support local populations as they take control of their own future. This level of funding cannot support a comprehensive response in most countries, but it can jump-start key programs when they are needed most, allowing time to seek alternative long-term funding. Furthermore, flexible spending authority would allow resources to be used to maximum effect. Consultation with Congress, backed by robust planning, would ensure accountability.

Fourth, S/CRS has established new management mechanisms that will foster interagency cooperation. At times when reconstruction efforts are judged central to U.S. interests, a senior interagency policy-level group, led by S/CRS and regional counterparts, will be created to make recommendations to the National Security Council, ensure that policy decisions are implemented, and secure quick action for serious problems. Tying all the key agencies into a common process will ensure that decisions lead to action, that military and civilian resources are fully utilized, and that information flows from the field to Washington.

Fifth, during U.S. or other military or peacekeeping operations, the new office will coordinate stabilization and reconstruction activities between civilian agencies and the military. As part of the military's planning effort, interagency civilian teams will deploy to regional combatant commands to develop strategies for stabilization and reconstruction. This type of involvement will help make certain that assumptions about civilian reconstruction capabilities remain realistic. After the planning stage, advance civilian teams will deploy with the military to help direct stabilization and reconstruction.

Finally, by creating a central locus for stabilization and reconstruction, the United States can cooperate more effectively with international partners. S/CRS has become a focal point within the U.S. government for relevant contacts with the UN, the European Union, regional organizations, and bilateral partners. With the Treasury Department and other State Department bureaus, S/CRS is also developing ongoing relationships with international financial institutions. And USAID plays a leading complementary role in coordinating with its development peers. Like the United States, other countries and institutions have recognized the need to build more effective programs to address conflict. Moreover, as the U.S. government becomes more adept at managing postconflict scenarios itself, it will be better able to coordinate its actions in the field with experts from around the world.

Global Rewards

If approved, the $124.1 million requested by the Bush administration in the 2006 budget would launch the conflict-response fund, initiate the development of a response corps, and provide the resources to train personnel and prepare operations for rapid deployment. Some may question why the United States should invest resources before crises emerge. But consider that a faster and more effective postconflict response could create better prospects for success and, in the case of U.S. military involvement, an earlier withdrawal of U.S. forces. For example, bringing one U.S. military division home from Iraq just one month early would save about $1.2 billion—and remove soldiers from harm's way.

The broader payoff is security. Today, stability requires more than maintaining a balance of power among strong states. Safety both here and abroad now depends on the ability of the United States and the international community to make sovereignty work—to establish democracies that improve the lives of ordinary individuals rather than of the ruling elite. The first step in this process must be to prevent conflict if possible, or to ensure a meaningful peace when conflict does occur. The world can do more to help those countries at risk of unrest or recovering from war. If successful, then over the longer term the United States will have enabled more people to enjoy the benefits of peace, democracy, and market economies. That can only be in everyone's best interest.

Chapter 10

Police and Homeland Security

By Gary W. Cordner and Kathryn E. Scarborough

Learning Objectives

- Identify the four systems in which police play a major role in the context of homeland security.
- Identify the elements of the mission of the Department of Homeland Security (DHS).
- Explain the term "interoperability" and why it is such a significant issue in policing today.
- Identify the four primary phases of the preparedness cycle, including the police role in each phase.
- Describe the National Incident Management System (NIMS) and its importance to policing today.
- Explain why information sharing is so important to the police role in homeland security.
- Describe intelligence-led policing and contrast it with community policing and problem-oriented policing.

The term "homeland security" was not common or familiar at the end of the twentieth century. At that time, just a few years ago, most experts, as well as casual observers, had no difficulty distinguishing between the *military* responsibility for homeland defense against our nation's enemies and the *police* responsibility for protecting people and property against crime. But then the world changed. We learned that modern international terrorism represents both a military threat and crime. Today, it is commonly accepted that the police have an important role to play in protecting both local communities and the nation against terrorism. Additionally, since Hurricane Katrina, it has become more widely accepted that the police role extends to "all hazards." The police role today within the context of homeland security includes important responsibilities related to crime, national defense, security, and safety.

The Police Role in Homeland Security

The police occupy a central role in homeland security, which is comprised of four important sectors or systems (see Figure 10.1). In accordance with the traditional view of policing, the police are the gatekeepers at the front end of the *criminal justice system*. They try to prevent crime, respond when crime does occur, and make criminal cases against suspected offenders. For the most part, the rest

Police and Terrorism

It has been said that 9/11 changed everything. This is certainly true for local police agencies and their chiefs. It is increasingly clear that federal agencies, such as the FBI and the U.S. Secret Service, can no longer work alone in protecting the United States from further attack. Rather, they must work in partnership with other public and private agencies, and most important, with local police. Local police can identify potential terrorists living or operating in their jurisdictions, they can help protect vulnerable targets, and they can coordinate the first response to terror attacks. These are heavy new responsibilities, but they cannot be shrugged off, because elected officials and the public will increasingly expect their police to be prepared.

Source: Graeme R. Newman and Ronald V. Clarke, *Policing Terrorism: An Executive's Guide* (Washington, DC: Office of Community Oriented Policing Services, 2008), Brief 01.

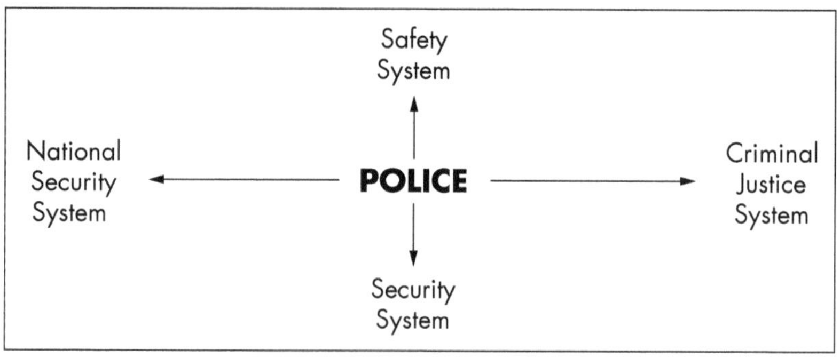

Figure 10.1 The Central Role of Police in Homeland Security

of the criminal justice system depends on the police, who control the entry point to the system and largely determine which offenders will be handled by prosecutors, judges, and correctional officials. What the new concept of homeland security has done is highlight several other systems in which the police also play important roles. Since the events of September 11, 2001, it is recognized that the police are an important part of the anti-terrorism effort, working in cooperation with the military, intelligence agencies, and others in the national defense or *national security system* (see "Police and Terrorism" above). America's 18,000 state, local, and special-purpose law enforcement agencies have much more intimate connections to local people and local communities than do the military, the CIA, or the FBI, and thus are a crucial resource for intelligence gathering and information sharing. Based on their own observations as well as information provided by the public, police can play the role of "first preventers" of terrorism.[1] Also, should terrorist acts occur, state and local police will be the first responders, not federal law enforcement agencies or the military. The crucial role of local police in the United States was highlighted by Israeli terrorism expert Tal Hanan:

> … local police officers must be trained regarding the threat [of suicide attacks], in particular the identification of suspicious activities and patterns used by terrorist cells in preparation for attacks. Police intelligence functions should include training on terrorist-related intelligence gathering methods, especially the monitoring of 'risk groups' and the components used to prepare bombs …. local police departments must ensure that every police officer is properly prepared for his or her role in combating suicide attacks, as they are the last line of defense.[2]

Since the experience of Hurricane Katrina, there is also greater recognition of the vital role that police play within the *safety system*. The key players in this system include fire protection personnel, emergency medical staff, public health, hospitals, and emergency managers. But often the very first responders to a serious emergency are police. Also, if a catastrophic situation requires mandatory evacuation, the police are the ones who will have to coordinate the evacuation and, if necessary, enforce it. Likewise, if a chemical, biological, or pandemic situation creates the need for quarantine, the police will be called upon to enforce that. In fact, through traffic safety and related functions the police have long contributed significantly to the safety system—it just has not been widely recognized or emphasized until recently.

One more system in which police play an important part is the *security system*. We traditionally think in terms of private security and expect to make a clear distinction vis-à-vis public police. However, private security and private policing have grown substantially in recent decades, and now often perform their duties in quasi-public places such as schools, malls, and even downtown business districts. Also, formal and informal public-private partnerships between private security and public police have become extremely common.[3] This takes on great importance in the homeland security context because it is estimated that 85 percent of the nation's critical infrastructure is privately owned.

Among these expanded police roles, perhaps the most challenging and controversial is the anti-terrorism role associated with national security. According to Larry Hoover, a new vision of American policing's national security role and responsibility has emerged.[4] He identifies four challenges faced by local law enforcement in participating in efforts addressing homeland security: technological, logistical, political, and ethical. The most significant technological challenge is *interoperability*. Interoperability is the ability to exchange information between organizations. Interoperability, or the lack thereof, pertains to both voice and data communications. Organizations that should be able to exchange information may include not only local, state, and federal law enforcement, but also public health departments, emergency medical services (ambulances and hospitals), and fire departments.

The most problematic logistical issue is getting data off written reports and out of officers' heads and into electronic formats that can be searched and shared. The necessity of this function is apparent, but if a majority of police officers' time is spent collecting data and entering it into a computer, little time is left for other responsibilities. Political issues take a variety of forms that focus on finances, role expectations, secrecy, and productivity, among others. Finally, ethical challenges include the issues of profiling and open records legislation. While technological, logistical, political, and ethical challenges are not new for the police administrator, those associated with homeland security are unlike ones faced before.

Contemporary policing themes of community engagement, collaboration, and globalization (discussed in Chapter 15) are all associated with the expanded homeland security mission. Community engagement highlights the need to design and implement roles for the public to play other than as victims of terrorism and disasters. Collaboration and coordination among so many separate law enforcement agencies, other providers of safety and security, and multiple levels of government is a huge challenge. In regard to community engagement and collaboration, the adoption of community policing since the 1980s certainly broadened the police perspective and taught the importance of partnerships, but the new homeland security mission and context present even greater demands and difficulties. The effect of globalization is that, increasingly, we may be dealing with issues that originate in another country, but affect us locally. This was demonstrated most dramatically by 9/11, but is also evidenced on a daily basis by transnational crime, especially cybercrime, that can realistically be regarded as a threat in every community in the nation.

Two critical events that occurred early in this twenty-first century have had a very significant impact on American society, including American policing and police administration. These two events were the terrorist attacks of September 11, 2001, and Hurricane Katrina in August 2005.

September 11, 2001

On September 11, 2001, 19 hijackers commandeered four airplanes leaving from Boston, Newark, and Washington, D.C., evading airport security at every juncture.[5] More than 3,000 people were killed in New York City at the World Trade Center, in Arlington, Virginia, just outside Washington, D.C. at the Pentagon, and in Stoney Creek Township, Pennsylvania, in an open field. Subsequently, it was learned that several of the hijackers had come to the attention of law enforcement authorities in the months and years leading up to the attack, but no one had "connected the dots." This gave rise to considerably more emphasis on intelligence analysis and information sharing. In addition, the heroic efforts of police, firefighters, and other emergency personnel in New York and at the Pentagon were hindered by overloaded communications networks and a lack of interoperability; that is, the inability of first responders from different agencies to communicate with each other across voice radio and data systems. This gave rise to technological enhancements and, just as important, improved protocols for shared and coordinated communications in emergencies.

To provide a description and analysis of the events of September 11, including preparedness and response, a 10-member commission, the National Commission on Terrorist Attacks Upon the United States (Public Law 107-306) was formed. The 9/11 Commission, as it was better known, eventually produced 41 recommendations that directed the revamping of the national security apparatus to combat terrorism and defend the nation against weapons of mass destruction. Several of the commissioners subsequently tracked the implementation of the commission's recommendations. In a status report in December 2005, the government received only one A grade, for making "significant strides in using terrorism finance as an intelligence tool."[6] Categories in which the government performed the worst included recommendations related to airline passenger screening, critical infrastructure assessment, and efforts to prevent terrorists from acquiring weapons of mass destruction; for 15 of the 41 recommendations, the government received either a D or F grade. Clearly, many components of terrorism preparedness and response needed more than just a quick fix.

Hurricane Katrina

On August 29, 2005, Hurricane Katrina struck Louisiana, Mississippi, Alabama, Florida, and Georgia with more than 1,300 fatalities, with additional people missing and many injured. Hurricane Katrina was the costliest hurricane, estimated at approximately $75 billion, and the fifth deadliest hurricane ever recorded in the United States. As if the hurricane itself was not enough devastation for the city of New Orleans, a breach of the levees that surrounded the city occurred and left thousands homeless. While much debate has ensued about whether the federal government, Louisiana state government, or New Orleans city government was most negligent (with particular attention focused on FEMA, the Federal Emergency Management Agency, a branch of the Department of Homeland Security), one thing is certainly evident: "the country was not prepared for a disaster of this magnitude, and the ensuing lack of planning, coordination, communication and response proved to be too little, too late."[7]

Hurricane Katrina: Breakdown in Law Enforcement

General unrest and lawlessness arose in crowded areas where people were uncertain about their survival, or rescue, or prospects for evacuation. In some areas, the collapse or absence of law enforcement exacerbated the level of lawlessness and violence. Several police departments lost dispatch and communication capabilities, police vehicles, administrative functions such as booking, and jails to confine arrested suspects. Tremendous additional burdens were imposed on the police, like search and rescue operations, that took priority over normal police functions. The extent of crime and lawlessness is difficult to determine, partly because of the loss of police record keeping during the disaster and partly because of unsubstantiated reporting by the media.

The breakdown of law enforcement was particularly notable in New Orleans. Despite the well-known threat from flooding, the New Orleans Police Department had not taken basic steps to protect its resources and ensure continuity of operations. For example, communications nodes, evidence rooms, and even emergency generators were housed in lower floors susceptible to flooding. When the levees broke and the floodwaters overtook police headquarters and district offices, the department lost its command and control and communications functions. Police vehicles believed to be moved out of harm's way were lost to the floodwaters. Hundreds of New Orleans Police Department officers went missing—some for legitimate reasons and some not—at the time they were needed most. This left the city unable to provide enough manpower and other resources to maintain law and order at shelters and on the streets.

Source: Select Bipartisan Committee to Investigate the Preparation for and Response to Hurricane Katrina, *A Failure of Initiative* (Washington, DC: U.S. Government Printing Office, 2006), p. 241.

A bipartisan congressional committee that investigated the problems that occurred during and after Hurricane Katrina devoted one of its chapters to law enforcement. They made several specific findings:

- A variety of conditions led to lawlessness and violence in hurricane-stricken areas.
- The New Orleans Police Department was ill-prepared for continuity of operations and lost almost all effectiveness.
- Lack of a government public communications strategy and media hype of violence exacerbated public concerns and further delayed relief.
- The Emergency Management Assistance Compact (EMAC) and military assistance were critical for restoring law and order.
- Federal law enforcement agencies were also critical to restoring law and order and coordinating activities.[8]

The congressional committee also found that law enforcement and public safety communications interoperability was once again, as it had been on September 11, 2001, a serious impediment to disaster response and recovery operations.

Department of Homeland Security

Since September 11, 2001, the U.S. has come to view terrorism and its responses to terrorism differently. The acts that occurred that day were unlike any we had ever seen before. While previous incidents at the World Trade Center in 1993 and in Oklahoma City in 1995 gave us pause, we had nothing of this magnitude—which affected so many people and professions—with which to compare these events. Similarly, Hurricane Katrina gave Americans a sober lesson about the severity of natural disasters and the critical importance of a coordinated and effective government response to emergency situations.

The Homeland Security Bill, enacted in 2003, mandated that 22 previously existing federal agencies with combined budgets of about $40 billion and employing 170,000 workers, be collapsed into one new cabinet-level department, with additional new divisions being created to fill existing gaps. This represented the most sweeping federal government reorganization since the creation of the Department

Homeland Security Presidential Directives (HSPD)

Homeland Security Presidential Directives are issued by the President on matters pertaining to Homeland Security.

- HSPD – 1: Organization and Operation of the Homeland Security Council.
- HSPD – 4: National Strategy to Combat Weapons of Mass Destruction.
- HSPD – 5: Management of Domestic Incidents.
- HSPD – 6: Integration and Use of Screening Information.
- HSPD – 7: Critical Infrastructure Identification, Prioritization, and Protection.
- HSPD – 8: National Preparedness.
- HSPD – 8: Annex 1: National Planning.
- HSPD – 9: Defense of United States Agriculture and Food.
- HSPD – 10: Biodefense for the 21st Century.
- HSPD – 11: Comprehensive Terrorist-Related Screening Procedures.
- HSPD – 12: Policy for a Common Identification Standard for Federal Employees and Contractors.
- HSPD – 13: Maritime Security Policy.
- HSPD – 15: U.S. Strategy and Policy in the War on Terror.
- HSPD – 16: Aviation Strategy.
- HSPD – 17: Nuclear Materials Information Program.
- HSPD – 18: Medical Countermeasures against Weapons of Mass Destruction.
- HSPD – 19: Combating Terrorist Use of Explosives in the United States.
- HSPD – 20: National Continuity Policy.
- HSPD – 20: Annex A: Continuity Planning.
- HSPD – 21: Public Health and Medical Preparedness.
- HSPD – 23: National Cyber Security Initiative.
- HSPD – 24: Biometrics for Identification and Screening to Enhance National Security.

Source: Department of Homeland Security. Homeland Security Presidential Directives. Available online at: http://www.dhs.gov/xabout/laws/editorial_0607.shtm.

of Defense in 1947. The Federal Bureau of Investigation (FBI) and the Central Intelligence Agency (CIA), two premier intelligence-gathering agencies, remained independent from the new Department of Homeland Security (DHS). However, many other agencies with law enforcement, border protection, emergency management, and intelligencegathering capabilities became part of DHS.

The initial mission of the new department was to: (1) prevent terrorist attacks, (2) reduce vulnerability to terrorism, and (3) minimize the damage from attacks that do occur.[9] The official mission statement was:

> We will lead the unified national effort to secure America. We will prevent and deter terrorist attacks and protect against and respond to threats and hazards to the nation. We will ensure safe and secure borders, welcome legal immigrants and visitors, and promote the free-flow of commerce.

Four divisions were created within the department, including Border and Transportation Security; Emergency Preparedness and Response; Chemical, Biological, Radiological, and Nuclear Countermeasures; and Information Analysis and Infrastructure Protection. Although these divisions represented the department's primary focus, DHS included numerous other functions that are not directly related to homeland security. For example, it is responsible for marine search and rescue, immigration and naturalization, counterfeiting, and drug smuggling.

The DHS is comprised of: (1) agencies that were already in existence prior to September 11, 2001; (2) new components that were created within the DHS, such as Customs and Border Protection (CBP), comprising elements of the preexisting U.S. Customs Service and U.S. Border Patrol; and (3) totally new agencies, such as the Transportation Safety Administration (TSA). Among the existing agencies that were moved to the new Department of Homeland Security were the Air Marshals, Border Patrol, Customs Service, Coast Guard, Federal Protective Service, Federal Emergency Management Agency, and Secret Service.

Other significant developments associated with the post-9/11 period and the creation of DHS were the USA PATRIOT Act (Uniting and Strengthening America By Providing Appropriate Tools Required to Intercept and Obstruct Terrorism, Public Law No. 107-56), the development of the National Incident Management System (NIMS), and the promulgation of a number of Homeland Security Presidential Directives (HSPDs). To-date, 24 of these HSPDs have been issued (see "Homeland Security Presidential Directives" on page 422). Not long after the DHS was created, a Six-Point Agenda was established to help in reaching the goals of the new federal agency. The Six-Point Agenda included the following:

1. Increase overall preparedness, particularly for catastrophic events.
2. Create better transportation security systems to move people and cargo more securely and efficiently.
3. Strengthen border security and border enforcement and reform immigration processes.
4. Enhance information sharing with our partners.

5. Improve DHS financial management, human resource development, procurement, and information technology.
6. Realign the DHS organization to maximize mission performance.

What did the creation of the Department of Homeland Security mean for local law enforcement? Following the formation of the DHS in 2003, we have seen a tremendous increase on the demands of federal, state, and local policing in the United States. The FBI shifted its focus from more traditional law enforcement activities, primarily reacting to criminal acts such as bank robberies, missing persons, corporate fraud, and other federal offenses, to focus more on domestic terrorism concerns and proactive investigations related to terrorism, transnational crime, and national security. State police agencies assumed some of the duties vacated by changing federal law enforcement priorities, and also became important components of state-level homeland security bureaus and intelligence fusion centers.[10] Local police agencies assumed greater responsibility for traditional personal and property crimes once referred to federal agencies, plus became responsible for increased duties protecting critical infrastructure, special events, and dignitaries.

The Preparedness Cycle

One of the best ways to think about the police role in homeland security is in reference to the domestic preparedness cycle that includes prevention, protection, response, and recovery. Preparedness emphasizes planning, training, and exercising to increase awareness, performance, and overall capabilities in all four phases of the preparedness cycle. It includes such activities as developing intelligence-collection and information-sharing protocols, developing local response plans, training first responders,

Homeland Security

Standard 46.3.1 The agency maintains liaison with other organizations for the exchange of information relating to terrorism.

Standard 46.3.2 A written directive establishes procedures for reporting and relaying terrorism-related intelligence/information to the proper task force or agency.

Standard 46.3.3 The agency provides terrorism awareness information within its service area.

Standard 46.3.4 The agency provides awareness level training for events involving hazardous materials.

Source: *Standards for Law Enforcement Agencies: The Standards Manual of the Law Enforcement Agency Accreditation Program*, Fifth Edition. Fairfax, VA: Commission on Accreditation for Law Enforcement Agencies, Inc., 2006.

identification of critical resources, and the development of mutual aid agreements both within and outside of jurisdictions.

Prevention

The prevention phase of the preparedness cycle applies readily to terrorism and transnational crime, but it also applies to natural disasters. Because terrorism and crime are intentional acts committed by people, it is well within reason to hope to prevent some of those acts from occurring. Natural disasters, however, such as hurricanes, tornados, and earthquakes, are not preventable in the same way (although it is certainly possible that future scientific developments might introduce some prevention measures). For these big natural disasters it is more practical to think in terms of preventing or reducing harm as opposed to actually preventing the disasters from occurring. Fewer lives might be lost and less property destroyed if people take the proper precautions, if buildings are upgraded, and so forth.

Information Sharing and Analysis Centers (ISAC)

The mission of the Information Sharing and Analysis Centers Council (ISAC Council) is to advance the physical and cyber security of the critical infrastructures of North America by establishing and maintaining a framework for valuable interaction between and among the ISACs and with government.

ISAC Council members include:

- Communications
- Electricity sector
- Emergency management and response
- Financial services
- Highway
- Information technology
- Multi-state
- Public transit
- Surface transportation
- Supply chain
- Water
- Research and education
- Maritime and research

Source: ISAC Council. *About the Council.* Available online at: http://www.isaccouncil. org/about/.

In some circles, this aspect of prevention is called *mitigation*. Regardless of terminology, the line between real prevention and mitigation of losses is often fuzzy. For example, we might not be able to prevent heavy rain that causes rivers and streams to flood. But with floodwalls and dams we can sometimes prevent the flooding, by channeling and controlling all the extra water brought by the rain. Similarly, by discouraging building in flood-prone areas, we reduce the impact of floods when they do occur. These activities clearly prevent flooding and/or losses from flooding, even though they do not prevent heavy rain.

Another important element of "all hazards," and one that is definitely an important target for prevention, is technological/industrial accidents.[11] These kinds of accidents have the potential to cause catastrophic losses. In the transportation sector, plane crashes, railway crashes, and ship/ferry sinkings can each account for hundreds of deaths per incident. Biological, chemical, and nuclear accidents have the potential for even larger-scale consequences.

The police play a secondary role in prevention or mitigation of natural disasters and technological catastrophes—engineers, scientists, air traffic controllers, airplane pilots, ship captains, risk managers, and government regulators play the primary role. For prevention of terrorism and serious crime, though, the police role is pivotal. Specific activities include border control, regulation of weapons and tools, patrol, investigation, surveillance, information and intelligence sharing, and community engagement. Involving the community in preventive activity is nothing new to the police. Citizens have historically provided the police with valuable information, which in some cases has become actionable intelligence.

Additionally, in a more formal role, the Intelligence Community (IC) ideally aids the police with the provision of information and intelligence. The IC is primarily composed of agencies of the national government, such as the CIA, FBI, National Security Agency, and military. Historically, however, the IC has emphasized sharing information and intelligence horizontally (across federal agencies), not vertically with state and local agencies, and this was a major criticism of the intelligence community following September 11, 2001. The government was graded low in the intelligence category, and still has not resolved many crucial issues concerning information sharing, especially vertical sharing of information and intelligence so that state and local agencies receive the same valuable information and intelligence that federal members of the IC receive. Later in this chapter, the section on "Information Sharing" discusses this vital aspect of terrorism prevention in more detail.

Protection

Protection is related to prevention but focused more on specific facilities, locations, and systems with the intent to make them less susceptible to attack and/or damage from disasters and accidents. The federal government identified 17 critical infrastructure sectors following the events of 9/11 and has focused its efforts on improving the overall security and protection of these different sectors. This is especially evident at the state and local levels, where police agencies have been called upon during heightened security alerts to increase patrols and presence at various facilities in their communities.

Despite these efforts, according to a report by the Government Accountability Office (GAO), the nation's critical infrastructures and key resources, including cyber and physical assets that have been designated essential to national security, national economic security, and national public health and safety, continue to be vulnerable to a wide variety of threats.[12] These post-9/11 vulnerabilities were again magnified following the damage from Hurricane Katrina, when the Gulf Coast experienced damage to oil platforms, pipelines, and refineries; water mains; electrical power lines; and cellular phone towers. The chaos resulting from this infrastructure damage disrupted the function of government and business and produced cascading effects beyond the Gulf Coast.

Damage or threats to critical infrastructure can be varied in method and extensiveness. The 9/11 attacks had a crippling effect on the airline industry. In 2004, authorities discovered detailed surveillance of the New York Stock Exchange and the Citigroup Center in the laptop computer of an Al Qaeda operative captured in Pakistan, part of a plan to target financial institutions in New York. The bombings of London's public transportation system demonstrated how an attack on a transportation system could disrupt a city's transportation and mobile telecommunications infrastructure. The 2008 terrorist attacks in Mumbai, India, targeted popular hotels.

Due to the fact that the private sector owns approximately 85 percent of the nation's critical infrastructure—such as banking and financial institutions, telecommunications networks, energy production, and transmission facilities—there is greater emphasis now for public and private sectors to form effective partnerships in order to protect these critical infrastructures. One new feature of infrastructure protection is a series of Information Sharing and Analysis Centers (ISAC), which are public-private partnerships designed to provide members of specific sectors with information about physical and cyber threats and vulnerabilities in order to increase protection levels (see "Information Sharing and Analysis Centers" on page 427).

Private ownership of so much critical infrastructure makes the significance of public-private partnerships readily apparent. Nowhere is this more evident than when considering the protection of border states and states with coastlines. In addition to terrorist threats, these states routinely deal with immigration, drug smuggling, cargo theft, and human trafficking challenges. Seaports are particularly challenging to protect because they are typically large, vast quantities of merchandise and material pass through them, they are a combination of land and sea, there is a mix of public and private ownership of property and facilities, and there is overlapping local, state, and federal jurisdiction. Commander Stephen Flynn (U.S. Coast Guard, ret.) points to the following weaknesses of our seaports and the challenges that face federal, state and local law enforcement:

> (1) Seaports cannot be separated from the international transport system to which they belong. Ports are in essence nodes in a network where cargo is loaded on or unloaded from one mode—a ship—to or from other modes—trucks, trains, and, on occasion, planes. Therefore, seaport security must always be pursued against the context of transportation security.
>
> (2) Port security initiatives must be developed within a regional and international context. Unilateral efforts to tighten security within U.S. ports without commensurate efforts to

improve security in the ports of our neighbors will lead shipping companies and importers to "port-shop"; i.e., to move their business to other market-entry points where their goods are cleared more quickly. Thus the result of unilateral, stepped-up security within U.S. ports could well be to erode the competitive position of important American ports while the locus of the security risk simply shifts outside of our reach to Canada, Mexico, or the Caribbean to ports such as Halifax, Montreal, Vancouver, and Freeport.

(3) Since U.S. ports are among America's most critical infrastructure, they should not be viewed as a primary line of defense in an effort to protect the U.S. homeland. The last place we should be looking to intercept a ship or container that has been co-opted by terrorists is in a busy, congested, and commercially vital seaport.[13]

Overview of NIMS

What NIMS Is:	What NIMS Is Not:
• A comprehensive, nationwide, systematic approach to incident management, including the Incident Command System, Multiagency Coordination Systems, and Public Information	• A response plan
• A set of preparedness concepts and principles for all hazards	• Only used during large-scale incidents
• Essential principles for a common operating picture and interoperability of communications and information management	• A communications plan
• Standardized resource management procedures that enable coordination among different jurisdictions or organizations	• Only applicable to certain emergency management/ incident response personnel
• Scalable, so it may be used for all incidents (from day-to-day to large-scale)	• Only the Incident Command System or an organization chart
• A dynamic system that promotes ongoing management and maintenance	• A static system

Source: *National Incident Management System* (Washington, DC: Department of Homeland Security, 2008), page 6. Available online at: http://www.fema.gov/pdf/emergency/ nims/NIMS_core.pdf.

Protecting critical infrastructure is a responsibility that is widely shared. Police do not bear all the responsibility for this—they have an important role to play, but so do private companies, private security, and non-police elements of the homeland security system. This situation of shared responsibility, together with the high degree of interdependence among various sectors of the economy and government, calls for a much greater level of coordination and collaboration than in the past.[14]

Response

Police are on duty and available 24 hours a day, seven days a week, 365 days a year. Often they are the only "first responders" working, so they are charged with responding to a wide variety of incidents, even those outside the realm of their regular duties. Historically, police have responded to various situations that were not necessarily law enforcement-related, but were considered part of the safety and service functions of policing. Despite being accustomed to this "jack of all trades" role, things changed dramatically for police (and other first responders) on September 11, 2001, when airplanes full of innocent people were used to assault high-rise office buildings and high-profile government installations. The essential nature of close collaboration between public agencies, as well as public-private collaboration, became obvious. This collective effort requires much more planning and coordination and, probably most important, communication, to operate effectively and safely in the post-9/11 world.

National Response Framework (NRF). The National Response Framework is intended to work in concert with the National Incident Management System. It requires that all jurisdictions establish a comprehensive, all-hazards approach to enhance the ability of the United States to manage disastrous and catastrophic incidents. The framework, which superseded the earlier National Response Plan, incorporates best practices and procedures from incident management disciplines—homeland security,

Police Role in Recovery

A large attack will place immediate demands on any police department; coping with the permanent changes that follow the attack can be equally demanding. At the World Trade Center disaster, police were called on for search and rescue operations. During the mitigation phase, police directed traffic, managed roadblocks, and supervised increased security in tunnels and bridges. Still others recorded and classified body parts recovered from the site, a job that went on for months. The aftereffects of the attack were so substantial that they necessitated the permanent redeployment of many police. The traffic flow in downtown Manhattan has been redesigned to cope with cleanup, reconstruction, and sightseeing. Deployment of police to maintain surveillance of possible targets (bridges, tunnels, subways, train stations, etc.) has changed the way the New York Police Department operates, probably permanently.

Source: Graeme R. Newman and Ronald V. Clarke, *Policing Terrorism: An Executive's Guide* (Washington, DC: Office of Community Oriented Policing Services, 2008), Brief 49.

emergency management, law enforcement, firefighting, public works, public health, responder and recovery worker health and safety, emergency medical services, and the private sector—and integrates them into a unified structure.[15] It forms the basis of how the federal government coordinates with state, local, and tribal governments and the private sector during major incidents.

The NRF identifies police, fire, public health and medical, emergency management, and other personnel as responsible for incident management at the local level. The idea is that an incident should be handled at the lowest possible organizational and jurisdictional level; however, if an incident expands beyond the capacity of the local response then the NRF provides a "seamless" integration with the federal government and a timely federal response to catastrophic incidents. These types of incidents are defined as high-impact, low-probability events, including natural disasters and terrorist attacks that result in extraordinary levels of mass casualties, damage, or disruption severely affecting the population, infrastructure, environment, economy, national morale, and/or government functions.

National Incident Management System (NIMS). The value of Incident Command Systems (ICS) is very evident to most police agencies that have ever had to deal with a hostage crisis or other life-threatening event that required a coordinated response and approach. Even before 9/11 and Hurricane Katrina, the importance of a coordinated and multiagency command and management system was something that fundamentally was understood at the local, state, and federal levels. The practice of Incident Command was formalized in the 1970s by a group of local, state and federal fire agencies following the brushfire season in Southern California. This group was referred to as Firescope (firefighting resources of California), which is still in existence today. Initially, ICS were more commonly the purview of the fire and emergency medical services, because they respond more often to major incidents with multiple personnel and multiple units. Law enforcement work more often involves just one or two officers, so the need for formalized incident command methods arises less often. One of the first documented and successful uses of formalized ICS for policing was during the 1984 Los Angeles Olympics.

It was not until 1996, when Executive Order No. 26 was signed, that a National Interagency Incident Management System-Incident Command System was required at the state level and would be required to be used during emergency operations. The idea was to create a common set of terms and procedures that provided for multiagency coordination, unified command, and the identification of equipment, resources, and common protocols for the handling of events and dissemination of information regarding an incident.

In 2003, HSPD #5 called for the creation of a comprehensive national approach to incident management. This initiative was referred to as the National Incident Management System (NIMS):

> ... a systematic, proactive approach to guide departments and agencies at all levels of government, nongovernmental organizations, and the private sector to work seamlessly to prevent, protect against, respond to, recover from, and mitigate the effects of incidents, regardless of cause, size, location, or complexity, in order to reduce the loss of life and property and harm to the environment. NIMS works hand in hand with the National Response Framework (NRF). NIMS provides the template for the management of incidents, while the NRF provides the structure and mechanisms for national-level policy for incident management.[16]

The most immediate implication for policing has been the requirement to take a number of ICS courses. The first series of courses had to be completed in 2006 and included ICS-100 (Introduction to Incident Command System) and ICS-200 (Single Resources and Initial Action Incidents). All federal, state, territorial, local, tribal, private sector, and nongovernmental personnel at the entry level, first-line supervisor level, middle-management level, and command and general staff level of emergency management operations were to complete these two courses. There was also a requirement in 2006 for all personnel with a direct role in emergency preparedness, incident management or response to complete IS-700 and IS-800 training. The IS-700 training is an awareness-level course that explains NIMS components, concepts, and principles, while the IS-800 course is also an awareness-level course and reviews the National Response Framework. These federally mandated training requirements were viewed as burdensome by many in the police field, although most recognized the need for better coordination of emergency responses to critical incidents (see "Overview of NIMS" on page 431).

The most controversial NIMS guidance given to police agencies by the federal government is an effort to standardize the language and rules of emergency response. Historically, local and state law enforcement officials have used "10-codes" in their radio communications. The criticism of that system is that there are no standard 10-codes. For example in Virginia, a "10–50" refers to a traffic accident, but in nearby Maryland, the same code indicates that an officer is in trouble. Currently many of the emergency responders within the same jurisdiction do not use a common 10-codes system, so the police code system is different from the fire department, which in turn uses different codes from the emergency medical services. Consequently, the Federal Emergency Management Agency (FEMA) issued a directive in 2006 with a series of new procedures, one of which was the phasing out of 10-codes (also called "brevity codes") and replacing them with phrases like "I'm at an accident scene" and other standard language. FEMA is insisting that the plain-language instructions apply for everyday use as well as national emergencies and is tying federal DHS funding to compliance.

Recovery

The police role in recovery from terrorism and disasters is secondary. Primary responsibility for community and economic recovery falls on FEMA, the Red Cross, other charitable organizations, insurance companies, private businesses, state and local governments, and individuals. However, as noted in "Police Role in Recovery" below, police may have to enforce curfews, regulate traffic, and protect facilities for weeks or even months after a major catastrophe, until other institutions and community dynamics return to some semblance of normalcy. The 9/11 and Katrina experiences illustrate how long it can take for this recovery period to progress to a point where the burden on police services eases.

A very important aspect of recovery for police departments, and a specific responsibility of police executives and police planners, is *continuity of operations*. The community counts on the police to provide services and protection in the days and months following a disaster or terrorist attack—what if police headquarters is flooded, cell phone service is knocked out, or the supply of gasoline is interrupted? This also applies during and following a pandemic situation, when a significant number of police personnel may be sick and quarantined.[17] Police departments need to develop contingency plans and back-up systems to ensure that they can continue to operate and protect the community during the post-disaster recovery period.

Technology and Training

Police activities in support of the all-hazards homeland security mission may require additional and/or enhanced equipment and training. In the first few years of the Department of Homeland Security, federal funding was made available to law enforcement agencies and other public safety agencies directly from Washington and through state homeland security offices to purchase specialized equipment. After making initial purchases of protective equipment to be used if chemical, biological, radiological, or nuclear (CBRN) material is encountered, many police agencies began to adopt two important principles that have subsequently guided homeland security technology procurement: (1) as much as possible, items should have dual purposes—that is, they should have everyday uses as well as emergency uses; and (2) regional cooperation should guide purchasing of the most expensive and specialized equipment. This latter point recognizes that every single police department does not need to own, and cannot afford to purchase and maintain, the most exotic and specialized technology, as long as they can get access to it when it is needed.

A national survey distributed in 2004 asked police officers and police administrators to identify their most critical technology needs related to the all-hazards homeland security mission. Interestingly, neither communications technology nor body armor was rated near the top of the most critical needs.[18] Respondents indicated that these technologies were very important, of course, but they also considered the equipment that was currently available to be satisfactory. The top five most critical technology needs according to the survey respondents were as follows:

Law Enforcement Administrators	**Law Enforcement Officers**
CBRN detection	CBRN detection
Explosives/firearms detection	Biohazard protection
Crime scene/crime lab	Explosives/firearms detection
In-vehicle/personal computers	Emergency database integration
Personal protective equipment	Vulnerability and threat assessment

These most critical technology needs seem to reflect serious concerns about safety and about the new and somewhat unfamiliar threats introduced by terrorism. Pre-9/11, when the potential threat from CBRN was primarily associated with industrial or transportation accidents, the police tended to define it as a hazardous materials (HazMat) issue that would be handled primarily by fire departments and emergency managers. When the same types of materials are introduced by terrorists, however, we call

Police Role in Recovery

- Passage of the USA Patriot Act in 2001 expanding (and/or clarifying) some law enforcement and intelligence authorities related to counterterrorism.
- Creation of U.S. Northern Command (NORTHCOM) in 2002.
- Publication of a National Strategy for Homeland Security in 2002.
- Creation of the Joint Regional Information Exchange System (JRIES) in 2002.
- Creation of the Department of Homeland Security (DHS) in 2002 (operational in 2003).
- Publication of a National Criminal Intelligence Sharing Plan in 2003.
- Creation of the Criminal Intelligence Coordinating Council (CICC) in 2004.
- Creation of the Homeland Security Information Network (HSIN) to replace JRIES in 2004.
- Creation of the Office of Director of National Intelligence (ODNI or DNI) in 2004.
- Creation of the National Counterterrorism Center (NCTC) in 2004.
- Establishment of 100 Joint Terrorism Task Forces (JTTFs) around the country, headed by the FBI with active participation from local and state police as well as other federal agencies, as of 2004.
- Passage of the Intelligence Reform and Terrorism Prevention Act establishing an Information Sharing Environment (ISE) in 2004.
- Publication of an updated National Strategy for Homeland Security in 2007.
- Publication of a National Strategy for Information Sharing in 2007.
- Passage of the 9/11 Commission Act establishing the Interagency Threat Assessment and Coordination Group (ITACG) in 2007.
- Publication of an Intelligence Community Information Sharing Strategy in 2008.
- Creation of 58 state and local information fusion centers as of 2008.

Source: Gary Cordner and Kathryn Scarborough, *Mapping the Terrain of Police, Military, and National Intelligence*. Report submitted to the Homeland Security Defense Education Consortium, May 2008.

them weapons of mass destruction (WMD) and there is a greater expectation that police will need to intervene. Police now have a heightened recognition of this responsibility, and are more aware of their equipment and technology needs to carry out the mission safely and successfully.

The all-hazards homeland security mission also raises new training requirements. As part of the National Preparedness Goal, DHS established a Target Capabilities List (TCL) that identifies what emergency responders and others will need to do in the event of various types of disasters and related scenarios.[19] Some of these target capabilities, such as medical surge or public health epidemiological investigation and laboratory testing, fall well outside the police mission. Others, such as intelligence fusion and analysis, public safety and security response, and citizen protection clearly involve the police. The Office of Grants and Training within DHS funds the development and delivery of homeland security-related training for police and others, and has set up a training portal at www.firstrespondertraining.gov to help meet the training needs of the entire nation.

A national survey administered in 2006 sought to identify the most serious training needs of police, other emergency responders, health workers, and general government officials related to the all-hazards homeland security mission.[20] The survey concentrated on the degree to which respondents had already received training related to each of the DHS target capabilities versus those TCLs that still presented substantial training needs. For each of 24 of the target capabilities, more than 50 percent of responding police agencies indicated that they would need training during the upcoming two years. The top five reported areas of training need for police agencies are noted below:

Percent of Agencies Indicating Training Need within Next Two Years

77%—planning for terrorism events
74%—information gathering and threat recognition
74%—prevention of cyber/computer crimes
73%—citizen protection: evacuation and/or in-place protection
71%—intelligence/information sharing and dissemination

Other types of training with high levels of reported need included interoperable emergency communications, critical infrastructure protection, responder safety and health, intelligence analysis and production, investigation of terrorism/WMD events, and citizen preparedness and participation. Clearly, the all-hazards homeland security mission has presented police with some new and complex responsibilities for which new and different types of training are needed.

Information Sharing

Effective homeland security depends heavily on information sharing. Preventing terrorism requires horizontal and vertical sharing of information and intelligence. Protection of critical infrastructure requires collaboration between public agencies and private companies, including sharing of confidential information. Emergency response to disasters, whether natural or man-made, requires coordination, including interoperable sharing of real-time information and databases. The critical importance of information sharing is difficult to overemphasize. Unfortunately, though, information and intelligence sharing within the law enforcement community has traditionally been inadequate, while the flow of information between law enforcement agencies and the Intelligence Community (military and national intelligence agencies) has been nearly nonexistent. The latter was due mainly to widely differing missions pre-9/11 police were focused mainly on domestic and transnational crime, while IC agencies were focused mainly on threats to national security posed by other nations.

Local police agencies are generally predisposed to share information with each other, but several complications have historically interfered with good intentions. The biggest is simply the fragmentation of the U.S. police system—some local police departments have 100-plus other local police agencies within a one- or two-county radius, making even local information sharing very complicated. A

Intelligence-Led Policing

Intelligence-led policing is a collaborative philosophy that starts with information, gathered at all levels of the organization, that is analyzed to create useful intelligence and an improved understanding of the operational environment. This will assist leadership in making the best possible decisions with respect to crime control strategies, allocation of resources, and tactical operations. The adoption of ILP processes requires a concerted effort by all parties, including analysts, operators, and senior leaders. For analysts the key components of this process include the creation of tactical, operational, and strategic intelligence products that support immediate needs, promote situational awareness, and provide the foundation for longer-term planning. For operators it requires becoming both better data collectors and better consumers of intelligence related products. This means shifting from emphasizing post-event evidence collection to constantly gathering all relevant data and ensuring it is provided for entry into appropriate databases, as well as drawing from the intelligence analysts and relevant databases all the information that is needed to support ongoing operations. Finally, adopting the ILP process requires senior leadership to actively engage the analysts and the operators to ensure the leadership has a sufficient picture of the operating environment and that it can act to distribute resources according to conclusions and priorities drawn from this understanding.

Source: Joseph R. Fuentes, *New Jersey State Police Practical Guide to Intelligence-Led Policing* (New York: Manhattan Institute, 2006), page 3. Available online at: http://www. cpt-mi.org/pdf/NJPoliceGuide.pdf.

few regional information systems have been created (ARJIS in San Diego and CLEAR in Chicago, for example), and it is common in some areas to hold monthly or quarterly regional meetings of chiefs, investigators, and/or intelligence specialists. Still, in such a fragmented system, it is difficult to determine who to share your information or intelligence with (who needs it) and to determine who to contact in search of particular information or intelligence.

Local police have routinely shared basic information with state police, other state agencies, and some federal agencies, especially the FBI. This is because state agencies and the FBI maintain the official repositories and databases of fingerprint cards, criminal histories, wanted persons, stolen cars, and other stolen property. In addition, state police typically oversee registration systems for handgun permits and sex offenders, while state motor vehicle agencies maintain records related to driver licenses and vehicle registrations. Also, the Uniform Crime Reporting (UCR) system is run by the FBI, using a process that involves local agencies submitting their reports to their state police or some other state agency, which combines the information and forwards it on to the FBI.

While local police have traditionally shared these kinds of basic information with state police and federal agencies on a routine basis, sharing of other information and intelligence has not always been routine. Local police often complain about a one-way flow of information in their relations with both state and federal agencies–they provide information, but they never get anything back. Resentment builds up, leading to less information sharing by local agencies.

In more recent years, though (beginning well before 9/11), the FBI, DEA, and Secret Service all created local-area federal task forces on which local police routinely participated. These violent crime, major drugs, and electronic crimes task forces enhanced operational cooperation as well as information and intelligence sharing. Those local police agencies participating in task forces often used them as their principal channels for information sharing with federal law enforcement agencies. Also, more than 7,000 law enforcement agencies joined six regional (multi-state) intelligence networks established to facilitate sharing of criminal intelligence.[21] In addition, local, state, and federal agencies jointly operated 31 regional and border HIDTA (High Intensity Drug Trafficking Areas) that emphasized drug-related information sharing and intelligence production.[22]

Following the events of September 11, 2001, a substantial number of reforms and other changes pertinent to intelligence and information sharing have been implemented (see "Post-9/11 Information Sharing Initiatives" below). Not surprisingly, these changes and reforms have not immediately solved all of the information and intelligence sharing problems associated with terrorism and homeland security. Local and state police have consistently complained that so-called "national" plans were too often really "federal" plans–created without state and local input, not responsive to state and local needs, often asserting federal control, and often creating new and unfunded state and local mandates.[23]

While obstacles and challenges remain, it is also the case that significant changes have occurred. Within one or two years after 9/11, local and state police were reporting increased responsibilities related to homeland security, increased contacts with a range of federal agencies, better relationships with federal authorities, and more open and effective information sharing.[24] More JTTFs were created (increasing from 35 to 100) with more state and local participation, and federal participation within state and local intelligence fusion centers became common. The overall trend has undoubtedly been in the direction of more information sharing.

The post-9/11 focus on local police has mainly been on their role as "eyes and ears" in local communities throughout the nation. In this respect they are seen as very important collectors of information, of raw data, that can be fed into the intelligence process in order to help analysts and others "connect the dots." Officers have been given training and also checklists to help them identify and report suspicious behavior that might have some connection to terrorism.[25] Community policing is seen as an ideal local police strategy because it helps officers get to know their communities and builds trust, making it more likely that residents will share important information with the police.[26] But local police need to be more than just information collectors—they should also be "first preventers" who are most likely to be in a position to prevent a terrorist act, both by gathering information and by taking action, when appropriate. This first preventer role is paired with the more familiar "first responder" role to make a logical and meaningful package that demonstrates the synergy between effective crime reduction tactics and counterterrorism and encourages local police to take their counterterrorism role more seriously.[27]

Important "linking pins" in the new information sharing environment are state fusion centers. These are designed to connect the hundreds of local agencies in each state with DHS and other federal components of the Intelligence Community, such as the National Counterterrorism Center and the Interagency Threat Assessment Coordination Group. Each of these centers is designed to receive, collate, and analyze intelligence from throughout its state, creating useful state-wide intelligence products when possible, and also passing important intelligence on to federal agencies when appropriate. The fusion centers also serve as points of contact for federal intelligence agencies when they need to make inquiries or disseminate information to officials within a state. Some of these fusion centers are focused exclusively on terrorism, but most have taken an "all crimes" or "all hazards" approach in order to be of greatest value to law enforcement agencies and others within their states.

Obviously, police executives have a much more complicated information sharing environment facing them today than in the past. In order to provide the best possible protection for their communities, as well as contribute to national security, they need to make sure that their officers are properly prepared to fulfill the roles of intelligence collectors, first preventers, and first responders. Also, police executives need to build relationships with, and maintain open channels of communication with, various state and federal agencies, including fusion centers, state police, the FBI, and DHS. Various national and state systems and plans have been established since 9/11 to make this kind of information sharing more feasible, but ultimately it depends on smart policing at the local level and a spirit of collaboration among all the parties involved in homeland security.

Intelligence-Led Policing

Police attention to terrorism and homeland security has led to the development of intelligence-led policing. Intelligence-led policing (ILP) is "a collaborative enterprise based on improved intelligence operations and community-oriented policing (COP) and problem-solving (POP)."[28] The basic idea is straightforward—police activity should be guided by information and intelligence.[29] Patrol assignments, for example, should be based on the latest information about crime and other problems, as opposed to simply assigning patrol officers to standardized beats and telling them to "be careful out there." Similarly, specialized units should be deployed based on intelligence about crime, drugs, and gang activity. Most importantly, these deployments should be based on current information and intelligence, as up-to-the-minute as possible. If intelligence gathering and intelligence analysis are effective, then ILP has the capacity to help police work smarter (see "Intelligence-Led Policing" below).

Thus far, it seems that more U.S. police agencies are exploring intelligence-led policing than actually implementing it on a wide scale. One stumbling block is that some agencies think of ILP simply in terms of terrorism and homeland security, not recognizing its applicability to crime control. Also, full-scale adoption of ILP requires an effective intelligence collection system (including collection of community intelligence), a robust capacity for intelligence analysis, and a strong command and management system capable of using intelligence products on a day-to-day basis to drive resource deployment and operational tactics. As this description indicates, ILP requires full integration of crime analysis and problem analysis with intelligence analysis.[30] In addition, it requires a sophisticated understanding of how community engagement and problem solving fit into ILP—too often, police executives seem to think they have to choose between COP, POP, and ILP, when in reality they should be thinking about how best to integrate these strategies into the most effective overall approach for their agency and their community. It will be interesting to see how these misunderstandings and challenges are sorted out over the next few years.

Summary

This chapter has outlined the modern police role in light of the events of September 11, 2001, the subsequent creation of the Department of Homeland Security, and the devastating experience of Hurricane Katrina in 2005. The federal government, police departments, and the rest of the emergency response community still have a lot of hard work to do to become more prepared and more adept at all-hazards prevention, protection, response, and recovery. The new homeland security mission presents extra workload for police agencies, significant new responsibilities, and major challenges. The crucial significance of information sharing has come to the forefront, and a new police strategy, intelligence-led policing, has emerged. It is quite likely that police work and police administration will never be the same again.

Discussion Questions

1. Do you agree that the police role has changed significantly since September 11, 2001? If not, why not? If you agree, identify and describe three ways in which it has changed.
2. Why is vertical information sharing so much more challenging than horizontal information sharing in the counterterrorism/domestic security context?
3. Who (agencies, levels of government, others) do you think should have had primary responsibility for recovery efforts following Hurricane Katrina?
4. Why is interoperability so important? Why do you think it hasn't already been achieved?
5. Do you think Intelligence-Led Policing or Community Policing is the best model to use to achieve Homeland Security? To what degree are these two police strategies compatible or in conflict?

Notes

1. G.L. Kelling and W.J. Bratton, "Policing Terrorism," *Civic Bulletin* No. 43 (September 2006). New York: Manhattan Institute for Policy Research.
2. T. Hanan, "Suicide Terrorism: The Overlooked WMD," *Journal of Counterterrorism & Homeland Security International* 9,3 (2003): 56–61.
3. COPS Office, *National Policy Summit: Building Private Security/Public Policing Partnerships to Prevent and Respond to Terrorism and Public Disorder* (Washington, DC: Office of Community Oriented Policing Services, 2005).
4. L.T. Hoover, "The Challenges to Local Police Participation in the Homeland Security Effort." *Subject to Debate* 16, 10 (October 2002): 1, 2, 4, 8–11.

5. National Commission on Terrorist Attacks Upon the United States, *The 9/11 Commission Report* (New York: W.W. Norton and Company, 2004).
6. 9/11 Public Discourse Project, "Final Report on 9/11 Commission Recommendations" (2005). Available online at http://www.9-11pdp.org/press/2005-12-05_report.pdf.
7. R.H. Ward, K.L. Kiernan, and D. Mabrey, *Homeland Security: An Introduction* (Newark, NJ: LexisNexis/Matthew Bender, 2006), p. 232.
8. Select Bipartisan Committee to Investigate the Preparation for and Response to Hurricane Katrina, *A Failure of Initiative* (Washington, DC: U.S. Government Printing Office, 2006), pp. 241–266.
9. Information on DHS in this section is drawn from R. White and K. Collins (eds.), *The United States Department of Homeland Security: An Overview* (Boston, MA: Pearson Custom Publishing, 2006).
10. C. Foster and G. Cordner, *The Impact of Terrorism on State Law Enforcement: Adjusting to New Roles and Changing Conditions* (Lexington, KY: Council of State Governments, 2005), available online at: http://www.csg.org/pubs/Documents/Misc0506Terrorism.pdf.
11. W.M. Evan and M. Manion, *Minding the Machines: Preventing Technological Disasters* (Upper Saddle River, NJ: Prentice Hall, 2002).
12. Government Accounting Office, *Critical Infrastructure: Challenges Remain in Protecting Key Sectors*, GAO-07-626T (Washington, DC: author, March 20, 2007).
13. See http://www.cfr.org/publication/4625/potential_impact_of_homeland_security_reorganization_on_federal_law_enforcement_activities.html?breadcrumb=%2Fbios%2F3301%2Fstephen_e_flynn%3Fpage%3D3.
14. K.A. Brown, *Critical Path: A Brief History of Critical Infrastructure Protection in the United States* (Fairfax, VA: Spectrum Publishing, 2006).
15. See http://www.dhs.gov/xprepresp/committees/editorial_0566.shtm.
16. *National Incident Management System* (Washington, DC: Department of Homeland Security, 2008), page 1. Available online at: http://www.fema.gov/pdf/emergency/nims/NIMS_core.pdf.
17. E.P. Richards, K.C. Rathbun, C.S. Brito, and A. Luna, *The Role of Law Enforcement in Public Health Emergencies: Special Considerations for an All-Hazards Approach* (Washington, DC: Bureau of Justice Assistance, 2006), p. 5.
18. G. Cordner, *Homeland Security Technology Survey Report* (Washington, DC: Public Safety and Security Institute for Technology, 2006).
19. See http://www.ojp.usdoj.gov/odp/docs/InterimNationalPreparednessGoal_03-31-05_1.pdf, p. 7.
20. G. Cordner, A. Cordner, and L. Mayberry, *Assessing the Needs of Rural Emergency Responders: National Training Needs Assessment 2006* (Richmond, KY: Rural Domestic Preparedness Consortium, 2007).
21. See http://www.riss.net/
22. Office of National Drug Control Policy. *High Intensity Drug Trafficking Areas* (2008). Available online at: http://www.whitehousedrugpolicy.gov/HIDTA/.
23. *Protecting Your Community From Terrorism: Improving Local-Federal Partnerships* (Washington, DC: Police Executive Research Forum, 2003); From Hometown Security to Homeland Security (Alexandria, VA: International Association of Chiefs of Police, 2005).
24. Foster and Cordner, *The Impact of Terrorism on State Law Enforcement: Adjusting to New Roles and Changing Conditions; Protecting Your Community From Terrorism: and The Production and Sharing of Intelligence* (Washington, DC: Police Executive Research Forum, 2005).

25. J. Meyer, "LAPD Leads the Way in Local Counter-Terrorism," *Los Angeles Times* (April 14, 2008); Bureau of Justice Assistance. *State and Local Anti-Terrorism Training Program* (2008). Information online at: https://www.slatt.org/Default.aspx.
26. *Local Law Enforcement's Role in Preventing and Responding to Terrorism* (Washington, DC: Police Executive Research Forum, 2001); M. Scheider and R. Chapman, "Community Policing and Terrorism," *Journal of Homeland Security* (April 2003). Available online at: http://www.homelandsecurity.org/newjournal/articles/scheider-chapman.html.
27. Kelling and Bratton, "Policing Terrorism."
28. M. Peterson, *Intelligence-Led Policing: The New Intelligence Architecture* (Washington, DC: Bureau of Justice Assistance, 2005).
29. J. Fuentes, *New Jersey State Police Practical Guide to Intelligence-Led Policing* (New York: Manhattan Institute, 2006). Available online at: http://www.cpt-mi.org/pdf/NJPo-liceGuide.pdf.
30. J.H. Ratcliffe, *Integrated Intelligence and Crime Analysis: Enhanced Information Management for Law Enforcement Leaders* (Washington, DC: Police Foundation, 2007).

Suggested Reading

Connors, Edward. *Planning and Managing Security for Major Special Events: Guidelines for Law Enforcement.* Washington, DC: Office of Community Oriented Policing Services, 2007.

Murphy, Gerard R. and Martha R. Plotkin. *Protecting Your Community From Terrorism: Strategies for Local Law Enforcement, Volume 1, Local-Federal Partnerships.* Washington, DC: Police Executive Research Forum, 2003.

Newman, Graeme R. and Ronald V. Clarke. *Policing Terrorism: An Executive's Guide.* Washington, DC: Office of Community Oriented Policing Services, 2008.

Ratcliffe, Jerry H. *Intelligence-Led Policing.* Devon, UK: Willan Publishing, 2008.

Richards, Edward P., Katherine C. Rathbun, Corina Sole Brito, and Andrea Luna. *The Role of Law Enforcement in Public Health Emergencies: Special Considerations for an All-Hazards Approach.* Washington, DC: Bureau of Justice Assistance, 2006.

Chapter 11

Department of Homeland Security: Summary and Cyber Needs

By Jennifer Michelle Stevens and Mark M. Lanier

After the devastating terror acts on September 11, 2001, the defensive umbrella of the United States was exposed as being inadequate for the protection of the homeland. Whenever a war is waged, utilizing the previous war's technology and strategy failure and defeat is a very likely outcome. History is full of examples whereby a dominant military power has been defeated by new tactics (e.g., the British in the American Revolutionary War, America in Vietnam). The next conflict is likely to be waged remotely by computers. Virtually every modern military weapons system, logistical system, and intelligence mission is predicated on computer technology. Many modern military planes cannot even fly without computers. In our view, computers, then, are the modern military's Achilles heel. This conclusion reviews the history and mission of Homeland Security. We conclude with a synopsis of what future cyber threats governments would be prudent to guard against.

After the successful September 11 attacks, President George W. Bush needed to quickly make a decision as to what to do about these foreign attacks and how to prevent or defend against future attacks. However, the president shares responsibility with Congress for the management and oversight of the federal bureaucracy, and this may constrain the president from making rapid decisions and being able to implement them in a timely fashion. Congress delays presidential powers because "departmental and agency reorganization requires legislative approval" (Lindsey and Daalder, 2003: 70). While this practice safeguards against tyranny, as the founding fathers envisioned, in this instance, President Bush wanted to immediately establish a small task force made specifically to protect the nation from future terrorist attacks. After the events of September 11, President Bush recognized that the nation was vulnerable to foreign attacks, and that an immediate change was absolutely necessary for the nation's safety. Bush was concerned that he did not have enough time to go through Congress and the traditional legislative procedures that come with creating a new government department. Also, if there was a massive governmental reorganization, there was the very real possibility that it would slow down efforts to coordinate federal antiterrorism programs (Arnold, 1998).

The options for a president to independently reorganize are limited to staff in the White House and the Executive Office of the President (EOP). In order to bypass the lengthy legislative process, Bush implemented an executive order. An executive order is one that directs White House staff, agencies, and/or departments to manage operations within the federal government and carries the same weight as laws passed by Congress. An executive order can also give the president the ability to create advisory posts in the EOP.

Executive Order: Office of Homeland Security

> *Today, dozens of federal departments and agencies, as well as state and local governments, have responsibilities affecting homeland security. These efforts must be coordinated at the highest level. So tonight I announce the creation of a Cabinet-level position reporting directly to me—the Office of Homeland Security.*
>
> —President George W. Bush

A week after the attacks of 9/11, President Bush implemented an executive order—the only tool available to him for swift action—to create the Office of Homeland Security (OHS). Bush stated this department was created to "lead, oversee, and coordinate a comprehensive national strategy to safeguard our country against terrorism and respond to any attacks that may come" (Bush, 2001). The OHS provided for a director to serve as an assistant to the president. Along with the Office of Homeland Security, a Homeland Security Council (HSC) was also created. The primary function of the OHS and HSC was to coordinate with each other. With the need for a director of the OHS, Bush appointed former governor of Pennsylvania Tom Ridge. Since Ridge was appointed at the president's discretion, Senate confirmation was not needed.

Problems with the Office of Homeland Security

Since the Office of Homeland Security was invested via executive order, this meant that the OHS held no permanent status. Also, it had a limited budget because the funds were discretionary and taken from the White House budget. Furthermore, the OHS lacked formal authority over other Cabinet departments and agencies, making it nearly impossible to gain respect and have coordination among other departments (Relyea, 2002: 401). The OHS was modeled after the National Safety Council (NSC), but the OHS did not follow the similar "culture of cooperation" that was obtained by the NSC (O'Hanlon, 2002: 99). It took the NSC nearly 50 years to achieve the kind of cooperation that is needed to operate at the department level.

The biggest problem faced by the OHS was the lack of statutory mandate in their structure. Had there been time for congressional legislation, the OHS would have been created as a permanent position in the Executive Office of the president, and would have brought the OHS into conformity with other important White House advisory positions (Daalder and Destler, 2001). There are numerous examples from the past that show how departments created via executive order have failed to achieve the duties and proposed mandate that were articulated for them. To provide an example, in 1943, President Franklin D. Roosevelt created the Office of War Mobilization (OWM) by executive order. A year later, Congress enacted the War Mobilization and Reconversion Act of 1994, in which they created a statuary Office of War Mobilization and Reconversion (OWMR), with the director subject to Senate confirmation. The OWMR no longer had operational responsibilities; it maintained a small staff tasked with coordinating a general policy framework, and ended up being extremely effective (Howard, pp. 402–404). Without the statutory change, the OWM did not stand a chance in creating the change that was required at the time.

Presidential Adviser Challenges

When former governor Tom Ridge was sworn in, many legislators did not approve the position of the director, or "presidential adviser," based on the fact that Ridge, like the OHS, had no statutory authority (Becker, p. A16). Legislators suggested to warn the president that Ridge's position was likely to be unsuccessful because of this lack of authority. For instance, with the lack of a congressional mandate, Ridge did not have an independent budget needed to carry out the duties of the OHS. Ridge, like all White House staff members, did not require Senate approval, and this shielded him from Congress (Howard, 2002: 401). However, without statutory authority, Ridge's "primary power for getting things done is the power of persuasion" (Balz and Allen, 2002: B1). His success was based solely on the cooperation and willingness of others to collaborate, cooperate, and respect his authority. Ridge's position was the focal point for harmonizing homeland security efforts that were scattered across innumerable agencies. A quote from Daalder and Destler explains Ridge's situation clearly:

> By chairing all interagency committees, Ridge and his office have the power to set the agenda, convene meetings, and forge consensus. But wielding that power effectively requires subtlety on Ridge's part. He needs to gain the cooperation of the many cabinet secretaries and agency directors … Neither Ridge nor anyone on his staff has the authority to tell others what to do—that must come from the acquiescence, if not support, of Ridge's peers themselves.

Ridge's first attempt of persuasion began in early 2002, when he tried to join together the Border Patrol, Coast Guard, and Customs Service into the department of Federal Border Administration. With little respect and limited support, the proposal was quickly dismissed. As mentioned, one of the biggest problems (other than the lack of statutory authority), was the absence of power allowing Ridge a budget. The presidential directive did not enable Ridge to formally certify the budget proposals of other entities with homeland security responsibilities that he was designated to coordinate (Cmar, 2002). Due to this, Ridge had authority and accountability issues that would have been resolved with congressional action.

What Could Have Been

Had Ridge been approved by the Senate, he would not have had constitutional struggles over testimonies, Ridge could have exercised powerful control over agency budgets dealing with homeland security, and he would not have had to fight for coordination among other departments (Cmar, 2002). If there was a statutory charter, it would have solved the problem of congressional accountability by making the directorship of the OHS subject to Senate confirmation. In this mandate, it would have included the authority for Ridge to certify agency budget submissions, which would have enabled Ridge to better coordinate counterterrorism programs. If Ridge had been able to obtain a substantive budget, this would have enabled him to prioritize and integrate agency proposals, with lesser dependence on his personal reputation, skills, and persuasion, although these attributes would have still been vital to Ridge's success.

United States Patriot Act

On October 26, 2002, President George W. Bush signed the Patriot Act. The Patriot Act oversees counterterrorism programs and deals with preventing future attacks. The main four points of the Patriot Act are:

- Allows investigators to use the tools that were already available to investigate organized crime and drug trafficking
- Facilitates information sharing and cooperation among government agencies so that they can better "connect the dots"
- Updates the law to reflect new technologies and new threats
- Increases the penalties for those who commit terrorist attacks

The Start of Reorganization

After the "presidential adviser" model proved to be insufficient for the task of coordinating governmental homeland security functions, proposals began in late 2001/early 2002 for a major governmental reorganization. President Bush tried to prevent a massive organization, but after the OHS failed, he was faced with two different options: create an independent agency, or create a 15th Cabinet-level department. Both would share the same objective: to build a more coherent approach to homeland security. Bush was aware that massive governmental reorganizations were costly and disruptive. Many critics claimed that the increased size and new layers of bureaucracy and coordination problems would outweigh the benefits of reorganization.

After careful review and consideration, in June of 2002, President Bush accepted the idea of creating a Cabinet-level department (Allen, 2002: A3). Critics were quick to point out that Bush had a total change of idea, moving from a non-statutory presidential advisory system to a full-fledged reorganization. Presidents promote reorganization of executive departments to reassert authority over the bureaucracy and/or signal their resolve to address urgent policy issues. Bush had decided to move ahead with a massive reorganization because the White House had recognized the widespread support for a significant reorganization, and Bush also wanted leverage over the structure of a new entity. Since a Cabinet-level department is established through legislation, Congress would now play a role in the new department. For their first step to reorganization of a new department, Congress and President Bush placed 170,000 federal employees under an "umbrella" department of Homeland Security (Tenpas and Hess, 2002: 581). On June 2002, Bush proposed joining 22 different agencies into a new Department of Homeland Security, divided into four divisions:

- Border and Transportation Security
- Emergency Preparedness and Response
- Chemical, Biological, Radiological, and Nuclear Countermeasures
- Information Analysis and Infrastructure Protection

This proposal left the Central Intelligence Agency (CIA) and the Federal Bureau of Investigation (FBI) outside of the new department (Curl, 2002: A1). Bush correctly noted that, "this is the most extensive reorganization of the federal government since the 1940's" (*New York Times*, 2002: A20).

First Problems of the Reorganization

Looking back over past reorganizations, presidents have been unsuccessful at large-scale reorganizations aimed at uniting similar functions under a single entity. A good example of a large-scale reorganization failing is that of President Richard Nixon's "superagencies" approach to domestic programs. President Lyndon Johnson attempted to create the Department of Economic affairs, but that failed; President Jimmy Carter also tried to create a Department of Natural Resources, but that failed as well (Nathan, 1975: Chs. 8 and 10). The first major problem of the new reorganization took place in the fall of 2002, when Congress recessed to campaign for their mid-term elections, putting the efforts of reorganization to a halt. Conflict between the executive and legislative branches of government started to develop, based on problems related to labor issues and presidential authority over employees in the new department. These problems resulted in veto threats from the Bush White House (Miller and Eilperin, 2002: 581). Also, the president and Congress were struggling to determine which functions to transfer to the new agency from the numerous agencies with responsibilities for counterterrorism. When these problems and other concerns surfaced, the departmental reorganization looked to be on the verge of another reorganizational failure. However, when the Republicans won the next congressional election, the bill was sent forth to Congress, which passed the bill in November 2002.

Homeland Security Act of 2002

The Homeland Security Act of 2002 was the legislative act responsible for establishing the Department of Security headed by the Secretary of Homeland Security. It establishes, in the Department of Homeland Security (DHS), the directorate for information analysis and infrastructure protection. In this act, it requires the secretary to establish procedures for sharing information and appointing a senior privacy officer to assume primary responsibility for privacy policy. The act establishes in the DHS an Office for Domestic Preparedness to prepare the United States for acts of terrorism. This act transfers many departments and agencies over to the DHS. The Secret Service and Coast Guard were both transferred over to the DHS. This act prohibits the secretary from substantially reducing the missions of the Coast Guard or its ability to perform them. It establishes within the Executive Office of the President a National Homeland Security to: asses United States objectives, commitments, and risks in the interest of homeland security; oversee and review federal homeland security policies; and offer recommendations to the president. Some of the major establishments created through this act are:

- The Department of Homeland Security
- Directorate for Information Analysis and Infrastructure Protection
- Critical Infrastructure Information Act of 2002
- Cyber Security Enhancement Act of 2002

The Homeland Security Act of 2002 holds many more regulations and responsibilities dealing with counterterrorism movements. The last major point of the Homeland Security Act is that it required the president to transmit a Department of Homeland Security reorganization plan to appropriate congressional committees within 60 days after the passing of the act.

Department of Homeland Security

After many controversies and problems with legislators and legislation, the Department of Homeland Security had finally been constructed. Bringing together 22 different agencies and departments, this department has been one of the biggest reorganizations in the government.

Learning from the Past and What Is to Come: Cyber Threats

Eight months before the attacks of 9/11, the Hart-Rudman Commission speculated, "a direct attack against American citizens on American soil is likely over the next quarter century" (Mufson, 2001:A2).

They advised the president that a congressionally mandated U.S. commission on national security was needed. The commission also released a report suggesting a massive overhaul of government foreign policy and security institutions. This commission made several important organizational recommendations, such as the creation of a National Homeland Security Agency with "responsibility for planning, coordinating, and integrating various United States agencies involved in homeland security" (Mufson, 2001: A2). They also brought up the idea of bringing together the Customs Service, Border Patrol, and Coast Guard to this new agency, but while "preserving them as distinct entities" (Wise, 2002). As stated earlier, Ridge had tried to accomplish this while being the director of OHS, but presumably due to his lack of authority and prestige, he could not get the proposal passed. Lastly, the Hart-Rudman Commission emphasized the need for homeland security to be integrated into a larger structure of national security intelligence. Parenthetically, the president did not understand the importance of these recommendations until the attacks of 9/11.

To date, the DHS has had both failure and success as a department. A success in averted attacks will never be quantified or publicized for security reasons. As the DHS keeps developing, the one thing that is yet to be completely addressed is the realm of cyber security. James Lewis, the director and senior fellow for the Center for Strategic and International Studies, states, "While DHS has improved, oversight for cyber security must move elsewhere" (Condon, 2008). Cyber crime has become a huge national problem, as have cyber terrorist threats. The GAO has been reporting on the cyber security efforts of the DHS and has tried to make recommendations for the department, but the department "still has not fully satisfied any of them" (*New York Times*, 2002: A39). It is unclear as to why the department is not where it should be with cyber security.

Currently, the undersecretary of the DHS is said to be handing the cyber security efforts, but there is rumor of a lack of leadership in the DHS in this area (Condon, 2008). It is unclear where cyber security would fall if moved out of the DHS, but it is clear that another change will need to be made to keep up with the constantly changing sphere of cyber crime. Cyber attacks pose the greatest threat to our national security, yet are seemingly not at the forefront of the DHS mission currently (though for security reasons, DHS efforts in this area would, by necessity, remain clandestine).

There are several challenges presented by cyber threats. Simon (2012) noted, "in today's cyber world, the ingredients necessary are available on a truly global scale and talent so common that children have successfully created cyber weapons" (pp. 31–32). Thus, the garage-and-basement hacker can pose a significant threat and operate independent of any government or terror organization. Another structural challenge is the fact that "governments are on the sideline in [the] main because they are too slow, too rigid, and too unimaginative to play in a realm dominated by individual talent, business imperatives, and rapid technological change" (Simon, 2012: 33). Finally, Simon astutely notes that "government budgetary practices also value maintenance and sustainability ahead of innovation and modernization" (2012: 33). For all these reasons, government in general, and the DHS in particular, are perhaps not well equipped or structured to address the numerous cyber threats facing contemporary society. Evidence of this is provided by the observation that in 2011, the profits generated by cyber crime surpassed the profits from the illegal drug trade (Simon, 2012: 32). The reasons for this change are numerous, and some are focused on government response.

Reasons for the proliferation of cyber crime are numerous and consequential for adequate response. For example, one problem that accompanies cyber crime is identification of the legal jurisdictions of the crime. Many times, cyber crime is not committed in the same state, or even the same country, as the victim or ultimate target. This leads to the law enforcement quandary of where the person should be prosecuted, convicted, and punished: in the location that attacker is in, or the location where the attack took place? Another problem that arises is that sufficient law does not address hacking or other cyber-related events. When an attacker is caught, the court has a hard time finding the right conviction mechanism (law) because cyber crime is a totally different terrain from what lies in the traditional system of law. The last major problem is that governments are not giving cyber crime the attention that is warranted. The government is well aware of this problem, but has yet to make any major efforts to combat cyber crime. The government has been hacked numerous times, and still has not tried any major changes to improve security. If the government is getting attacked easily, that leaves the people of the nation extremely vulnerable to attacks. With Homeland Security being the safeguard of the nation, it is a necessity for them to increasingly focus on cyber security to protect the homeland—which was the first priority implemented in the creation of Homeland Security. The DHS will therefore focus on cyber terror threats rather than other types of cyber crime.

Concentrating on cyber terror attacks requires attention to three critical areas: the perimeter, the inside, and awareness. All three are relevant and important. The perimeter refers to protecting the internal workings and operations of an agency or business from a cyber attack. The inside is where great vulnerabilities exist, and Simon (2012) considers this most serious. Awareness requires an absolute understanding of all the hardware and software used in the entity. Awareness, according to Simon, also requires "absolute knowledge of each aspect of all hardware and software, total configuration control, and, especially, complete knowledge of each component's base (normal) state" (2012: 43). This is necessary so that changes (perhaps an attack) to the normal state or anomalies can immediately be identified and isolated.

In conclusion, since computer ownership and Internet connectivity are easily accessible and affordable, computer and Internet usage have increased exponentially. In part, because of this increased usage and availability, computer crime has become an ever increasing social problem—and more importantly, a national security threat. At any given time, a person could be hacked by an individual or a cyber group of activists. Likewise government is vulnerable. Cyber crime is common in today's society, but many governments are having a hard time defending cyber attacks and predicting when attacks are going to take place. The government does not have enough intelligence to combat cyber crimes. Universities rarely have degrees in areas relating to the cyber world, which means that people are not equipped with the knowledge needed to defend cyber crime. Not only are people underqualified for positions in cyber security, the laws that are presently implemented do not always cover the terrain of cyber threats. Also, people have trouble getting help when dealing with cyber problems because most local and state law enforcement agencies are not trained to guard against, identify, or prosecute cyber crime. There are so many victims—worldwide—that it is nearly impossible for law enforcement to have an impact. As Simon stated, awareness is critical, and papers such as these begin to create awareness.

References

Allen, M. (2002, June 25). "Bush Campaigns for Creating Homeland Security Department," *Washington Post*, p. A3.

Arnold, P. (1998). *Making the managerial presidency: Comprehensive reorganization planning*, 1905–1996, 2nd ed. Lawrence: University Press of Kansas.

Balz, D., and Allen, M. (2002, May 22 and May 26). "Democrats Shift Focus on Probe of Terror Warnings; Look at Agencies' Role Sought as Strategists Urge Party to Challenge Bush on Domestic Issues, Not Sept. 11," *Washington Post*, p. A15; "The 97-Pound Weakling: For Ridge to Succeed He Needs Authority from Bush," *Newsday*, p. B1.

Becker, E. (2002, April 22 and June 9). "Bush Is Said to Consider New Security Department," *New York Times*, , p. A16; T. Oliphant, "Bush's Security Flip-Flop," *Boston Globe*, p. E7.

Bush, G. W. (2001, October 29). *HSPD-1: Organization and Operation of the Homeland Security Council* (United States). Retrieved May 5, 2013, from http://www.fas.org/irp/offdocs/nspd/hspd-1.htm

Condon, S. (2008, September 17). "Critics: Homeland Security unprepared for cyberthreats," CBS News. Retrieved May 29, 2013, from http://news.cnet.com/8301-13578_3-10043665-38.html

Cmar, T. (2002, summer). "Office of Homeland Security," *Harvard Journal on Legislation* 39.

Curl, J. (2002, June 7). "Bush Proposes New Security Agency; Asks Hill to Quickly Make Place in Cabinet," *Washington Times*, p. A1.

Daalder, I., and Destler, I. M. (2002, Summer) "Behind America's Front Lines: Organizing to Protect the Homeland" *Brookings*

Daalder, I., and Destler, I. M. (2001, November 7). "How NOT to Reorganize for Homeland Security," *Hill*, Retreived May 23, 2013 http://www.puaf.umd.edu/faculty/papers/destler/HowNotToReorganize.htm

Hart-Rudman Commission, (1993, January 20) "Road Map for National Security: Imperative for Change: The Phase III Report of the U.S. Commission on National Security/21st Century," p. 15. http://www.nssg.gov/PhaseIIIFR.pdf

Lindsay, J., and Daalder, I. (2003) "Whose Job Is It? Organizing the Federal Government for Homeland Security." In *American politics after September 11*, p. 70, James M. Lindsay, Ed. Cincinnati: Atomic Dog Publishers.

Miller, B., and Eilperin, J. (2002, September 23). "Obscure Labor Issues Block Homeland Security Bill," *Washington Post*, p. A8.

Mufson, S. (2001, February 1). "Overhaul of National Security Apparatus Urged; Commission Cites U.S. Vulnerability," *Washington Post*, p. A2.

Nathan, R. P. (1975). "The Plot That Failed: Nixon and the Administrative Presidency." New York: Wiley; on Johnson and Carter, see P. E. Arnold, *Making the managerial presidency*, chs. 8 and 10, respectively. (1998, July)

O'Hanlon, M., Orszag, P. R., Daalder, I. H., Destler, I. M., Gunter, D. L., Litan, R. E., and Steinberg, James. (2002). *Protecting the American Homeland: A Preliminary Analysis*, p. 99. Washington, DC: Brookings Institution; full report at http://www.brook.edu/dybdocroot/fp/projects/homeland/report.htm

Relyea, H. C. (2002, June). "Homeland Security: The Concept and the Presidential Coordination Office—First Assessment," *Presidential Studies Quarterly* 32, p. 401.

Relyea, H. C. (2004, April, 21) "Homeland Security: The Concept and the Presidential Coordination Office—First Assessment," pp. 402–404.

Rudman, Warren and Hart, Gary "Restructuring for Security,". *New York Times*, June 13, 2002, p. A39. Quoted in "Traces of Terror; The Plan: 'We Have Concluded That Our Government Must Be Reorganized,'" *New York Times*, June 7, 2002, p. A20.

Simon, J. M. (2012). "Toward a General Understanding of Cyber Safety." #CyberDoc No Borders: No Boundaries: National Doctrine of the Cyber Era. Sample, T. R., and Swetmen, M. S. (Eds.). Battelle: Arlington, VA: Potomac Institute Press.

Tenpas, K. D., and Hess, S. (2002). "The Bush White House: First Appraisals," *Presidential Studies Quarterly* 32, p. 581.

Wise, C. R. (2002, March/April). "Organizing for Homeland Security," *Public Administration Review* 62.

www.ingramcontent.com/pod-product-compliance
Ingram Content Group UK Ltd.
Pitfield, Milton Keynes, MK11 3LW, UK
UKHW051453290926
15471UKWH00008B/1175

9 781621 317128